Igor Chikalov

Average Time Complexity of Decision Trees

Intelligent Systems Reference Library, Volume 21

Editors-in-Chief

Prof. Janusz Kacprzyk
Systems Research Institute
Polish Academy of Sciences
ul. Newelska 6
01-447 Warsaw
Poland
E-mail: kacprzyk@ibspan.waw.pl

Prof. Lakhmi C. Jain
University of South Australia
Adelaide
Mawson Lakes Campus
South Australia 5095
Australia
E-mail: Lakhmi.jain@unisa.edu.au

Further volumes of this series can be found
on our homepage: springer.com

Vol. 5. George A. Anastassiou
*Intelligent Mathematics: Computational
Analysis*, 2010
ISBN 978-3-642-17097-3

Vol. 6. Ludmila Dymowa
Soft Computing in Economics and Finance,
2011
ISBN 978-3-642-17718-7

Vol. 7. Gerasimos G. Rigatos
*Modelling and Control for Intelligent
Industrial Systems*, 2011
ISBN 978-3-642-17874-0

Vol. 8. Edward H.Y. Lim, James N.K. Liu,
and
Raymond S.T. Lee
*Knowledge Seeker – Ontology Modelling for
Information
Search and Management*, 2011
ISBN 978-3-642-17915-0

Vol. 9. Menahem Friedman and Abraham
Kandel
Calculus Light, 2011
ISBN 978-3-642-17847-4

Vol. 10. Andreas Tolk and Lakhmi C. Jain
Intelligence-Based Systems Engineering, 2011
ISBN 978-3-642-17930-3

Vol. 11. Samuli Niiranen and Andre Ribeiro
(Eds.)
*Information Processing and Biological
Systems*, 2011
ISBN 978-3-642-19620-1

Vol. 12. Florin Gorunescu
Data Mining, 2011
ISBN 978-3-642-19720-8

Vol. 13. Witold Pedrycz and Shyi-Ming Chen
(Eds.)
Granular Computing and Intelligent Systems,
2011
ISBN 978-3-642-19819-9

Vol. 14. George A. Anastassiou and Oktay
Duman
*Towards Intelligent Modeling: Statistical
Approximation Theory*, 2011
ISBN 978-3-642-19825-0

Vol. 15. Antonino Freno and Edmondo
Trentin
Hybrid Random Fields, 2011
ISBN 978-3-642-20307-7

Vol. 16. Alexiei Dingli
*Knowledge Annotation: Making Implicit
Knowledge Explicit*, 2011
ISBN 978-3-642-20322-0

Vol. 17. Crina Grosan and Ajith Abraham
Intelligent Systems, 2011
ISBN 978-3-642-21003-7

Vol. 18. Achim Zielesny
From Curve Fitting to Machine Learning,
2011
ISBN 978-3-642-21279-6

Vol. 19. George A. Anastassiou
*Intelligent Systems: Approximation by
Artificial Neural Networks*, 2011
ISBN 978-3-642-21430-1

Vol. 20. Lech Polkowski
Approximate Reasoning by Parts, 2011
ISBN 978-3-642-22278-8

Vol. 21. Igor Chikalov
Average Time Complexity of Decision Trees,
2011
ISBN 978-3-642-22660-1

Igor Chikalov

Average Time Complexity
of Decision Trees

 Springer

Dr. Igor Chikalov
Mathematical and Computer Sciences
and Engineering Division
4700 King Abdullah University of Science
and Technology
Thuwal 23955-6900
Kingdom of Saudi Arabia
E-mail: igor.chikalov@kaust.edu.sa

ISBN 978-3-642-22660-1   e-ISBN 978-3-642-22661-8

DOI 10.1007/978-3-642-22661-8

Intelligent Systems Reference Library   ISSN 1868-4394

Library of Congress Control Number: 2011932437

© 2011 Springer-Verlag Berlin Heidelberg

Typeset & Cover Design: Scientific Publishing Services Pvt. Ltd., Chennai, India.

Printed on acid-free paper

9 8 7 6 5 4 3 2 1

springer.com

To my wife Julia and our daughter Svetlana, for always believing in me and loving me, no matter what.

Foreword

It is our great pleasure to welcome a new book "Average Time Complexity of Decision Trees" by Igor Chikalov. This book is devoted to the study of average time complexity (average depth and weighted average depth) of decision trees over finite and infinite sets of attributes. It contains exact and approximate algorithms for decision tree optimization, and bounds on minimum average time complexity of decision trees. The average time complexity measures can be used in searching for the minimum description length of induced data models. Hence, there exist relationships of the presented results with the minimum description length principle (MDL).

The considered applications include the study of average depth of decision trees for Boolean functions from closed classes, the comparison of results of the performance of greedy heuristics for average depth minimization with optimal decision trees constructed by dynamic programming algorithm, and optimization of decision trees for the corner point recognition problem from computer vision.

The book can be interesting for researchers working on time complexity of algorithms and specialists in machine learning.

The author, Igor Chikalov, received his PhD degree in 2002 from Nizhny Novgorod State University, Russia. During nine years he was working for Intel Corp. as a senior software engineer/research scientist in machine learning applications to the control and diagnostic problems of semiconductor manufacturing. Since 2009 he is a senior research scientist in King Abdullah University of Science and Technology, Saudi Arabia. His current research interests include supervised machine learning and extensions of dynamic programming to the optimization of decision trees and decision rules.

The author deserves the highest appreciation for his outstanding work.

Mikhail Moshkov

May 2011

Andrzej Skowron

Preface

The monograph is devoted to theoretical and experimental study of decision trees with a focus on minimizing the average time complexity. The study resulted in upper and lower bounds on the minimum average time complexity of decision trees for identification problems. Previously known bounds from information theory are extended to the case of identification problem with an arbitrary set of attributes. Some examples of identification problems are presented giving an evidence that the obtained bounds are close to unimprovable. In addition to universal bounds, we study effectiveness of representing several types of discrete functions in a form of decision trees. In particular, for each closed class of Boolean functions we obtained upper bounds on the average depth of decision trees implementing functions from this class.

The monograph also studies the problem of algorithm design for optimal decision tree construction. An algorithm based on dynamic programming is proposed that describes a set of optimal trees and allows for subsequent optimization on other criteria. Experimental results show applicability of the algorithm to real-life applications that are represented by decision tables containing dozens of attributes and several thousands of objects.

Beside individual identification problems, infinite classes of problems are considered. It describes necessary conditions on such classes in order to have polynomial complexity algorithms for optimal decision tree construction.

The presented results can be of interest for researchers in test theory, rough set theory and machine learning. Some results may be considered for including in graduate courses on discrete mathematics and computer science. The monograph can be used as a reference to prior results in the area.

Some results were obtained in collaboration with Dr. Mikhail Moshkov and published in joint papers [51, 52, 53, 54, 56]. I am heartily thankful to Dr. Moshkov for help in preparing this book.

I would like to acknowledge and extend my gratitude to Victor Eruhimov for fruitful discussions about applications of decision trees and Dr. Andrzej Skowron for constructive criticism and suggestions for improvement of the book.

Thuwal, Saudi Arabia,
April 2011   Igor Chikalov

Contents

1   Introduction 1
    1.1   Basic Notions 4
        1.1.1   Information Systems 4
        1.1.2   Problems Over Information Systems 4
        1.1.3   Decision Trees 5
        1.1.4   Decision Tables 5
        1.1.5   Complexity Measures of Decision Trees 6
    1.2   Overview of Results 7
        1.2.1   Bounds on Average Weighted Depth 7
        1.2.2   Representing Boolean Functions by Decision Trees 9
        1.2.3   Algorithms for Decision Tree Construction 11
        1.2.4   Restricted Information Systems 12

2   Bounds on Average Time Complexity of Decision Trees .. 15
    2.1   Known Bounds 16
    2.2   Bounds on Average Weighted Depth 16
    2.3   Upper Bound on Average Depth 18
        2.3.1   Process of Building Decision Trees $Y_{U,\Psi}$ 19
        2.3.2   Proofs of Theorems 2.3 and 2.4 20
    2.4   On Possibility of Problem Decomposition 26
        2.4.1   Proper Problem Decomposition 26
        2.4.2   Theorem of Decomposition 27
        2.4.3   Example of Decomposable Problem 37

3   Representing Boolean Functions by Decision Trees 41
    3.1   On Average Depth of Decision Trees Implementing Boolean
       Functions 42
        3.1.1   Auxiliary Notions 42

3.1.2   Bounds on Function $\mathcal{H}_B(n)$ 43

3.1.3   Proofs of Propositions 3.1-3.13 45

3.2   On Branching Programs with Minimum Average Depth 58

4   Algorithms for Decision Tree Construction 61

4.1   Algorithm $\mathcal{A}$ for Decision Tree Construction 62

4.1.1   Representation of Set of Irredundant Decision Trees .. 63

4.1.2   Procedure of Optimization......................... 66

4.2   Greedy Algorithms 69

4.3   Modeling Monotonic Boolean Functions by Decision Trees ... 72

4.4   Constructing Optimal Decision Trees for Corner Point

Detection.. 74

4.4.1   Corner Point Detection Problem 74

4.4.2   Experimental Results 76

5   Problems over Information Systems 79

5.1   On Bounds on Average Depth of Decision Trees Depending

Only on Entropy 79

5.2   Polynomiality Criterion for Algorithm $\mathcal{A}$ 83

A   Closed Classes of Boolean Functions 87

A.1   Some Definitions and Notation 87

A.2   Description of All Closed Classes of Boolean Functions 89

References.. 95

Index.. 101

Chapter 1
Introduction

Decision trees appeared in 50-60s of the last century in theoretical computer science [14, 64, 80] and applications [24, 37]. Similar objects are also considered by natural and social sciences, for example, taxonomy keys [30] or questionnaires [63]. Decision trees naturally represent identification and testing algorithms that specify the next test to perform based on the results of the previous tests. A number of particular formulations were generalized by Garey [27] as identification problem that is a problem of distinguishing objects described by a common set of attributes. More general formulation is provided by decision table framework [34, 65] where objects can have incomplete set of attributes and non-unique class labels. In that case, acquiring class label is enough to solve the problem: identifying a particular object is not required. In this context, decision trees found many applications in test theory [39, 45, 46, 81], fault diagnosis [14, 60, 72], rough set theory [61, 62], discrete optimization, non-procedural programming languages [34], analysis of algorithm complexity [38], computer vision [74], computational geometry [69].

Decision tree is also a way of representing data in a structured hierarchical manner. It describes a recursive partitioning of a set of objects into groups according to the attribute values. Such representation reveals various patterns in data like object similarity and common characteristics of several objects. If objects are divided into classes, decision tree gives an idea of which attributes are important for assigning an object to a certain class. In machine learning problems, decision trees show ability to generalization that is capturing strong dependencies only and ignoring the weak ones which are resulted from a finite sample size and do not reflect properties of the data source [8, 71]. Compact decision trees are easily interpreted by human experts that makes it favorable over other models. The state-of-the-art statistical modeling techniques like tree ensembles [7, 26] use decision trees for its insensitivity to outliers and

I. Chikalov: Average Time Complexity of Decision Trees, ISRL 21, pp. 1–14.
springerlink.com   © Springer-Verlag Berlin Heidelberg 2011

uniform way of dealing with numeric and categorical (discrete unordered) attributes.

In most cases, multiple decision trees are available for the same problem. Not all of them are equally favorable. Depending on the application, a tree is required to have minimal storage complexity or guaranteed time complexity in all cases, or minimal expected number of tests. This leads to different strategies for building of decision trees. Bounds on the minimum tree complexity and algorithms for optimal tree constriction are studied in test theory [21, 49, 46, 47, 81], rough set theory [57, 58, 79], search theory [1], machine learning [33, 71]. It was discovered, that for almost all criteria, the problem of building an optimal decision tree is NP-hard, and for many cases there are results preserving a polynomial time approximation. The problem of design of effective algorithms for building decision trees is still open. Though, recent advances proved that greedy algorithms [12, 32] build trees that are close to optimal for some cases.

In this monograph, several known results on the average time complexity of decision trees are generalized and a number of new problems are considered. The main goal is to obtain bounds on the minimum average time complexity of decision trees and design effective algorithms for building decision trees for some classes of information systems. Methods of combinatorics, probability theory and complexity theory are used in the proofs as well as concepts from various branches of discrete mathematics and computer science.

The monograph consists of five chapters. Chapter 2.4.3 considers bounds on the minimum average weighted depth of decision trees. Upper and lower bounds on the average time complexity of decision trees were known previously for a problem with a complete set of attributes. These bounds depending only on the entropy of probability distribution follow from results of coding theory [41, 77] and are widely applied in search theory (see, e.g. [1]). Chapter 2.4.3 generalizes these bounds to the case of the average weighed depth of decision trees for an arbitrary identification problem. In the first section, an upper bound on the average weighted depth of decision trees and more precise bound on the average depth are proved. These bounds depend on the entropy and a parameter $M(z)$, which is introduced by Moshkov in [46]. An analogous parameter of the exact learning problem is called *extended teaching dimension* [4, 33]. In general case, calculating $M(z)$ is computationally intractable, but for several classes of problems, either exact value or tight bounds on $M(z)$ can be obtained by theoretical analysis.

The second section describes conditions on the problem structure and the probability distribution for objects that enable problem decomposition. Under these conditions an optimal decision tree for the initial problem can be synthesized from optimal decision trees for simpler problems. This technique is used to build a class of problems for which the minimum average depth of decision tree is close to its upper bound given in the first section.

Chapter 3.2 is devoted to several applications that can effectively use decision trees. It consists of two sections. The first section studies the average depth of decision trees implementing Boolean functions. A Shannon type function is considered that describes growth of the average depth of decision trees with growth of the number of arguments in the functions being implemented. For each closed class of Boolean functions [68, 36], a lower and an upper bound is obtained on a Shannon type function characterizing this class. The obtained results are compared to the analogous results for the depth of decision trees described in [48]. The notion of decision table partition used in the proofs is similar to the notion of system of nonoverlapping coverings of Boolean cube used in spectral methods of digital logic [6], but the type of covering is estimated from the parent closed class of the function rather than its spectral properties. It allows to improve lower bounds on the average depth of decision trees for some functions (e.g., the voting function and the logical sum). The second section shows that each branching program with the minimum average weighted depth is a read-once branching program. Due to this fact, known exponential lower bounds on the number of nodes in read-once branching programs for several combinatorial problems [59, 66, 83, 84, 85] are applicable to branching programs with the minimum average weighted depth.

Chapter 4.4.2 is devoted to algorithms for decision tree construction. The first section describes an algorithm A that builds a set of decision trees with the minimum average weighted depth for a problem given in a form of decision table. The idea of the algorithm is based on dynamic programming [27, 42, 60, 76]. The second section describes experimental results of using A for implementing Boolean functions by decision trees. The third section is devoted to greedy algorithms. It describes a general scheme of greedy algorithm, defines several data impurity functions, and describes results of a comparative study of performance of several greedy algorithms applied to data sets from UCI Machine Learning Repository [25]. The fourth section describes results of applying A to a practical problem of computer vision—fast detection of corner points [75].

Chapter 5.2 considers a class of information systems called restricted information systems. It consists of two sections. The first section proves that for restricted information systems (and only for such systems), there exist upper bounds on the average depth of decision trees that depend only on the entropy of object probability distribution. The second section gives necessary and sufficient conditions that make the time complexity of the above considered algorithm A limited from above by a polynomial on the number of rows in a table. These conditions contain the requirement for the information system to be restricted. In [1], the average depth of decision tree is studied for some problems (e.g., the problem of finding a leak in a pipeline). The obtained results generalize bounds from [1] to an arbitrary restricted information system and give polynomial algorithms for building optimal decision trees.

The monograph contains mainly theoretical results that can be used for design of effective algorithms for building decision trees and for analysis of complexity of representing various objects by decision trees. These results can be of interest for researchers in test theory, rough set theory and logical analysis of data. The monograph can be used as a part of a course for graduate students and Ph. D. studies.

1.1   Basic Notions

Denote $\omega = \{0, 1, 2, \ldots\}$, and for $k \in \omega \setminus \{0, 1\}$, denote $E_k = \{0, \ldots, k - 1\}$.

1.1.1   Information Systems

Let A be a nonempty set, F a set of functions defined on A and taking values from E_k, so that for any $f \in F$, the condition $f \not\equiv \text{const}$ holds. The functions from F are called *attributes*, and the pair $U = (A, F)$ is called *k-valued information system* (or simply *information system*) .

A *weight function* for the information system U is a function of the form $\Psi : F \to \{1, 2, \ldots\}$ that assigns a *weight* $\Psi(f)$ to each attribute $f \in F$.

1.1.2   Problems Over Information Systems

A *problem over the information system* U is defined by a tuple $z = (\nu, f_1, \ldots, f_n)$, where $\nu : E_k^n \to \{0, 1, \ldots, k^n - 1\}$ and $f_1, \ldots, f_n \in F$. The problem z consists in finding the value $z(a) = \nu(f_1(a), \ldots, f_n(a))$ for an arbitrary element $a \in A$.

Two elements a and b from A are *equivalent for the problem* z if $f_i(a) = f_i(b)$ for $i = 1, \ldots, n$. This equivalence relation defines a partition of A into nonempty *equivalence classes* $Q_1, \ldots, Q_s$. Let us denote by T_z the set $\{\bar{d}_1, \ldots, \bar{d}_s\} \subseteq E_k^n$ where $\bar{d}_i = (f_1(a), \ldots, f_n(a))$ and $a \in Q_i$, $i = 1, \ldots, s$. A problem z is called *diagnostic* if for any two tuples $\bar{d}_i, \bar{d}_j \in T_z$, $\bar{d}_i \neq \bar{d}_j$, the condition $\nu(\bar{d}_i) \neq \nu(\bar{d}_j)$ holds.

Probability distribution for the problem z is a mapping $P : T_z \to \omega \setminus \{0\}$. For $\bar{d} \in T_z$, the value $P(\bar{d})/\sum_{\bar{\delta} \in T_z} P(\bar{\delta})$ can be interpreted as a probability of the event $(f_1(a), \ldots, f_n(a)) = \bar{d}$ for an arbitrary element a from A.

1.1.3   Decision Trees

A *decision tree for the problem* $z = (\nu, f_1, \ldots, f_n)$ is a finite oriented tree with root in which:

- each nonterminal node is assigned with an attribute from the set $\{f_1, \ldots, f_n\}$ (i.e. decision trees use only the attributes listed in the description of the problem z);
- each nonterminal node has exactly k outgoing edges which are labeled with the numbers $0, \ldots, k-1$ respectively;
- each terminal node is assigned with a number from ω.

Let us describe the algorithm represented by a decision tree. Let the input be an element $a \in A$. First, the root is assigned to be the current node. Let us describe one step of the algorithm. If the current node is terminal, the algorithm returns as result the number assigned to the current node and finishes. Otherwise, let f_c be the attribute assigned to the current node. For $\delta = 0, \ldots, k-1$, let e_δ be the edge that leaves the current node and is labeled with δ. The value $f_c(a)$ is calculated, and the node that the edge $e_{f_c(a)}$ enters becomes the current node. Then the algorithm proceeds to the next step.

1.1.4   Decision Tables

Let U be a k-valued information system, Ψ a weight function for U, $z = (\nu, f_1, \ldots, f_n)$ a problem over U, and P a probability distribution for z. Let $T_z = \{\bar{d}_1, \ldots, \bar{d}_s\}$. The set T_z can be represented as a rectangular table filled with numbers from E_k. Rows of the table correspond to the equivalence classes, columns to the attributes, and each number is the value of the corresponding attribute for all elements of the corresponding equivalence class. Let us assign the i-th column with the weight of the attribute f_i for $i = 1, \ldots, n$, and assign the row $\bar{d}_j$ with the numbers $\nu(\bar{d}_j)$ and $P(\bar{d}_j)$ for $j = 1, \ldots, s$. We will denote the resulted table T_z^* and call it *decision table for the problem* z. Further several algorithms will be considered that take as input a tabular representation of the problem z.

A two-player game can be associated with the table T_z^*. The first player thinks of a row $\bar{d}$ from T_z^*. The goal of the second player is to ascertain the number $\nu(\bar{d})$ assigned to the row $\bar{d}$ in T_z^*. The second player is allowed to ask questions of the following form: he can choose a column and ask which number is on the intersection of the column and the row that the first player has in mind. A strategy of the second player can be represented in a form of a decision tree.

Denote $\Omega_z = \{(f_i, \delta) : f_i \in \{f_1, \ldots, f_n\}, \delta \in E_k\}$, and denote Ω_z^* the set of all finite words in the alphabet Ω_z including the empty word λ. Let us extend the mapping Ψ to the set Ω_z^*. Let α be an arbitrary word from Ω_z^*. If $\alpha = \lambda$, then $\Psi(\alpha) = 0$. For $\alpha = (f_{i_1}, \delta_1) \ldots (f_{i_t}, \delta_t)$, $t > 0$, assume $\Psi(\alpha) = \Psi(f_{i_1}) + \ldots + \Psi(f_{i_t})$.

Let $\alpha \in \Omega_z^*$. Define a *separable subtable* $T_z \alpha$ of the table T_z in the following way. If $\alpha = \lambda$, then $T_z \alpha = T_z$. Let $\alpha \neq \lambda$ and $\alpha = (f_{i_1}, \delta_1) \ldots (f_{i_m}, \delta_m)$. Then $T_z \alpha$ is the subtable of the table T_z that contains only the rows which have the numbers $\delta_1, \ldots, \delta_m$ in the columns $i_1, \ldots, i_m$ respectively. We will say that a table is *terminal* if it contains no rows or $\nu(x) \equiv$ const on the set of rows. Denote $S(z)$ the set of all nonterminal separable subtables of the table T_z.

For an arbitrary table T from $S(z)$, we denote by $D(T)$ the number of rows in T and denote $N(T, P) = \sum_{\bar{d} \in T} P(\bar{d})$.

Let Γ be a decision tree for the problem z. Set to the correspondence to each path $\xi = v_1, r_1, \ldots, v_t, r_t, v_{t+1}$ in Γ a word $\pi(\xi) \in \Omega_z^*$. Let $t \geq 1$, for $j = 1, \ldots, t$, the node v_j be assigned with an attribute f_{i_j}, and the edge r_j, leaving v_j and entering v_{j+1} be assigned with a number δ_j. Then $\pi(\xi) = (f_{i_1}, \delta_1), \ldots, (f_{i_t}, \delta_t)$. We assume $\pi(\xi) = \lambda$ for a path ξ consisting of a single node .

A path from the root to a terminal node is called *complete*. Denote $\Xi(\Gamma)$ the set of complete paths in decision tree Γ. One can see that $\bigcup_{\xi \in \Xi(\Gamma)} T_z \pi(\xi) = T_z$, and for any two different complete paths ξ_1, ξ_2, the relation $T_z \pi(\xi_1) \cap T_z \pi(\xi_2) = \emptyset$ holds.

We will state that a *decision tree Γ solves the problem z* if for an arbitrary row $\bar{d} \in T_z$, the terminal node of the complete path ξ such that $\bar{d} \in T_z \pi(\xi)$ is assigned with the number $\nu(\bar{d})$. In other words, for an arbitrary element $a \in A$, the terminal node of the path on which computations for a are performed is labeled with the number $z(a)$.

1.1.5   Complexity Measures of Decision Trees

Let $U = (A, F)$ be an information system, Ψ a weight function for U and z a problem over U. Let Γ be a decision tree for z that solves the problem z. For an arbitrary row $\bar{d} \in T_z$, denote $\xi^{\bar{d}}$ the complete path in Γ on which computations for the n-tuple of attribute values $\bar{d}$ are performed.

As the main complexity measure the *average weighted depth of the decision tree Γ relative to the probability distribution P* (or, briefly, *P-average weighted depth of Γ*) will be used. It is defined in the following way:

$$h_\Psi(\Gamma, P, z) = \frac{1}{N(T_z, P)} \sum_{\bar{d} \in T_z} \Psi(\pi(\xi^{\bar{d}})) P(\bar{d}) .$$

In addition to the average weighted depth, the weighted depth will be used as a complexity measure of decision trees. *Weighted depth* of decision tree Γ is defined as follows:

$$g_\Psi(\Gamma, z) = \max_{\bar{d} \in T_z} \Psi(\pi(\xi^{\bar{d}})) .$$

If $\Psi \equiv 1$, then the considered complexity measures are called *average depth* and *depth*, and denoted $h(\Gamma, P, z)$ and $g(\Gamma, z)$. Further we will omit the symbol z in the notations $h(\Gamma, P, z)$ and $g(\Gamma, z)$ if it is clear which problem is meant.

Denote $h_\Psi(z, P)$ and $h(z, P)$ respectively the minimum P-average weighted depth and the minimum P-average depth of the decision tree for the problem z that solves z. For a weight function Ψ, a problem z and a probability distribution P, a decision tree that solves z and has the minimum P-average depth is called *optimal for z and P*, and a tree that solves z and has the minimum P-average weighted depth is called *optimal for Ψ, z and P*.

1.2   Overview of Results

This section briefly describes main theoretical results of the monograph.

1.2.1   Bounds on Average Weighted Depth

Let U be an information system and Ψ a weight function for U. First, we define a parameter $M_\Psi(z)$ for a problem $z = (\nu, f_1, \dots, f_n)$ over U. If $z(x) \equiv \text{const}$ on the set A, then $M_\Psi(z) = 0$. Otherwise, for an arbitrary tuple $\bar{\delta} = (\delta_1, \dots, \delta_n) \in E_k^n$, denote $M_\Psi(z, \bar{\delta})$ the minimum natural number m such that there exist numbers $i_1, \dots, i_r \in \{1, \dots, n\}$ possessing the following conditions: $\Psi(f_{i_1}) + \dots + \Psi(f_{i_r}) \leq m$ and either the set of solutions on A of the system of equations $\{f_{i_1}(x) = \delta_{i_1}, \dots, f_{i_r}(x) = \delta_{i_r}\}$ is empty or $z(x) \equiv \text{const}$ on this set. Then

$$M_\Psi(z) = \max_{\bar{\delta} \in E_k^n} M_\Psi(z, \bar{\delta}) .$$

If $\Psi \equiv 1$, then the parameter $M_\Psi(z)$ is denoted by $M(z)$.

As a parameter of probability distribution P we will use *the entropy of probability distribution*

$$H(P) = \log_2 N(T_z, P) - \frac{1}{N(T_z, P)} \sum_{\bar{d} \in T_z} P(\bar{d}) \log_2 P(\bar{d}) \ .$$

If a diagnostic problem contains all possible attributes, then a known noiseless coding theorem is applicable saying that the minimum average depth of decision tree is between $H(P)$ and $H(P) + 1$. The following theorem generalizes the lower bound to the case of the average weighted depth of decision tree for an arbitrary diagnostic problem.

Theorem. *(Theorem 2.2 from Sect. 2.2) Let U be a k-valued information system, Ψ a weight function for U, z a diagnostic problem over U, and P a probability distribution for z. Then*

$$h_\Psi(z, P) \geq \frac{H(P)}{\log_2 k} \ .$$

The following theorem gives an upper bound on the minimum average weighted depth of decision tree for an arbitrary problem.

Theorem. *(Theorem 2.3 from Sect. 2.2) Let U be an information system, Ψ a weight function for U, z a problem over U, and P a probability distribution for z. Then*

$$h_\Psi(z, P) \leq M_\Psi(z)(H(P) + 1) \ .$$

Since the average depth of decision tree is a particular case of the average weighted depth, the above considered bounds hold for the average depth as well. However, the upper bound on the average depth can be improved.

Theorem. *(Theorem 2.4 from Sect. 2.3) Let z be a problem over an information system U, and P a probability distribution for z. Then*

$$h(z, P) \leq \begin{cases} M(z) \ , & \text{if } M(z) \leq 1 \ , \\ M(z) + 2H(P) \ , & \text{if } 2 \leq M(z) \leq 3 \ , \\ M(z) + \frac{M(z)}{\log_2 M(z)} H(P) \ , & \text{if } M(z) \geq 4 \ . \end{cases}$$

In Sect. 2.4, a possibility of reduction is considered for a problem over 2-valued information system. An algorithm is described that constructs a decision tree for the initial problem from decision trees for subproblems that form so-called proper decomposition of the original problem. The section also contains sufficient conditions for the synthesized decision tree to be optimal relative to the average depth.

The decomposition technique allows finding decision trees with the minimum average depth for some classes of problems. In Sect. 2.4.3, it is used to prove that the upper bound on the average depth of decision tree given by Theorem 2.4 is close to unimprovable.

Theorem. *For an arbitrary natural numbers $m \geq 2$, $n \geq 3$ there exists an information system U_m^n, a problem z_m^n over U_m^n with m^n classes of equivalence and a probability distribution $P_m^n \equiv 1$ such that*

$$h(z, P) \geq \frac{(M(z) - 2)H(P)}{2 \log_2 M(z)} .$$

This theorem immediately follows from Theorem 2.6 given in Sect. 2.4.3.

1.2.2   Representing Boolean Functions by Decision Trees

In Chap. 3.2, efficiency of representation of Boolean functions by decision trees is studied. A Boolean function $f(x_1, \ldots, x_n)$ can be represented as a problem $z = (f, x_1, \ldots, x_n)$ over the information system $U_n = (E_2^n, \{x_1, \ldots, x_n\})$. The problem z has two equivalence classes Q_0 and Q_1 containing the sets of binary tuples on which f takes the values 0 and 1 respectively. A decision tree solving the problem z is called *a decision tree implementing f*. Denote by $g(f)$ and $h(f)$ respectively the minimum depth of a decision tree implementing f and the minimum average depth of a decision tree implementing f relative to the probability distribution $P \equiv 1$.

Denote by $\dim f$ the number of arguments of the function f. Let B be a set of Boolean functions. Consider the functions

$$\mathcal{G}_B(n) = \max\{g(f) : f \in B, \dim f \leq n\}$$

and

$$\mathcal{H}_B(n) = \max\{h(f) : f \in B, \dim f \leq n\}$$

that characterize the growth in the worst case of the minimum depth and the minimum average depth of decision trees implementing Boolean functions from B with growth of the number of function arguments. Note that $\mathcal{H}_B(n) \leq \mathcal{G}_B(n)$ for any n.

Section 3.1.2 contains several statements that give an upper and a lower bounds of $\mathcal{H}_B(n)$ for each closed class of Boolean functions B. The notation of closed classes of Boolean functions is in accordance with [36]; the classes and the class inclusion diagram are described in Appendix A. It is shown that the function $\mathcal{H}_B(n)$ is either limited from above by a constant or grows linearly. The work [48] gives exact values of $\mathcal{G}_B(n)$ for each closed class of Boolean functions. The following two theorems characterize the relation between $\mathcal{H}_B(n)$ and $\mathcal{G}_B(n)$.

Theorem. *(Theorem 3.2 from Sect. 3.1) Let B be a closed class of Boolean functions, and n a natural number. Let at least one of the following conditions hold:*

a) $n = 1$;
b) $B \in \{O_1, \ldots, O_9, L_1, \ldots, L_5, C_1, C_2, C_3\}$;
c) $B \in \{C_4, D_1, D_3\}$ and n is odd;
d) $B \in \{D_1, D_2, D_3\}$ and $n = 2$.

Then $\mathcal{H}_B(n) = \mathcal{G}_B(n)$. If none of the conditions a), b), c), d) hold, then $\mathcal{H}_B(n) < \mathcal{G}_B(n)$.

Theorem. *(Theorem 3.3 from Sect. 3.1) Let B be a closed class of Boolean functions. Then*

a) $\lim_{n \to \infty} \mathcal{H}_B(n)/\mathcal{G}_B(n) = 0$ if $B \in \{S_1, S_3, S_5, S_6, P_1, P_3, P_5, P_6\}$;
b) $\mathcal{H}_B(n)/\mathcal{G}_B(n) = 1$ if $B \in \{O_1, \ldots, O_9, L_1, \ldots, L_5, C_1, C_2, C_3\}$;
c) $\lim_{n \to \infty} \mathcal{H}_B(n)/\mathcal{G}_B(n) = 1$ if $B \in \{C_4, M_1, \ldots, M_4, D_1, D_2, D_3\}$;
d) $\lim_{n \to \infty} \mathcal{H}_B(n)/\mathcal{G}_B(n) = 1/2$ if $B \in \{F_1^\infty, \ldots, F_8^\infty\}$;
e) $1/2 - \varepsilon(n) < \mathcal{H}_B(n)/\mathcal{G}_B(n) < 1$ where $\varepsilon(n) = O(1/\sqrt{n})$ if $B \in \{F_1^\mu, \ldots, F_8^\mu\}$ and $\mu \geq 2$.

If the number of nodes is estimated in addition to the average weighted depth, it is reasonable to combine isomorphic subtrees in decision tree. The resulted object is called *branching program*. A branching program is called *read-once* if in any path from the root to a terminal node, each attribute encounters at most once. The following theorem shows that the requirement to a branching program to be read-once is rather strong as any branching program with the minimum average weighted depth is a read-once branching program.

Theorem. *(Theorem 3.4 from Sect. 3.2) Let U be a 2-valued information system, Ψ a weight function for U, z a problem over U, and P a probability distribution for z. Let G be a branching program for z that solves z and is optimal for Ψ, z and P. Then G is a read-once branching program.*

In [66], it is shown that a read-once branching program implementing the function $Mult : \{0, 1\}^{2n} \to \{0, 1\}$ (the middle bit of the multiplication of two n-bit integers) contains at least $2^{\Omega(\sqrt{n})}$ nodes. In [83, 84, 85], the function $n/2 - Clique - Only : \{0, 1\}^{n^2} \to \{0, 1\}$ is considered that receives adjacency matrix of a graph with n nodes and takes the value 1 if and only if the graph contains a $n/2$-clique and does not contain any other edges. It is shown that a read-once branching program implementing the function $n/2 - Clique - Only$ contains at least $2^{\Omega(n)}$ nodes, while there is a branching program with $O(n^3)$ nodes implementing $n/2 - Clique - Only$ such that any attribute appears at most twice in each path. In [59], it is shown that a branching program implementing the characteristic functions of Bose-Chaudhuri codes contains at least $exp(\Omega(\sqrt{n}/2))$ nodes.

Theorem 3.4 shows that the branching programs that are optimal relative to the average weighted depth have the same or greater number of nodes than the read-once branching programs with the minimum number of nodes.

1.2.3   Algorithms for Decision Tree Construction

Let $U = (A, F)$ be an information system and $z = (\nu, f_1, \ldots, f_n)$ a problem over U. Let T be a separable subtable of T_z. For $i \in \{1, \ldots, n\}$, denote $E(T, i)$ the set of numbers contained in i-th column of the table T, and denote $E(T) = \{i : i \in \{1, \ldots, n\}, |E(T, i)| \geq 2\}$.

Among decision trees for the problem z that solve z we distinguish *irredundant* decision trees. Consider an arbitrary node w of the tree Γ and denote $path(\Gamma, w)$ the path from the root to w. Let $T = T_z \pi(path(\Gamma, w))$ be a terminal subtable and $\nu(x) \equiv r$ on the set of rows of the table T for some $r \in \omega$. Then w is a terminal node labeled with r. Let T be a nonterminal subtable. Then w is labeled with an attribute f_i where $i \in E(T)$. Finally, each node w such that $T_z \pi(path(\Gamma, w)) = \emptyset$ is labeled with the number 0.

The following proposition shows that among irredundant decision trees, at least one has the minimum average depth.

Proposition. *(Proposition 4.1 from Sect. 4.1)   Let U be an information system, Ψ a weight function for U, z a problem over U, and P a probability distribution for z. Then there exists an irredundant decision tree that is optimal for Ψ, z and P.*

Denote by $Tree(T_z)$ the set of irredundant decision trees for the problem z. In Sect. 4.1, an algorithm $\mathcal{A}$ is described that constructs the set of optimal irredundant decision trees for the problem z. At the first stage of the algorithm, a graph $\Delta(z)$ of separable subtables of the table T_z is constructed. The graph in some sense describes all irredundant decision trees for the problem z. Then the algorithm reduces the graph $\Delta(z)$ resulting the graph $\Delta_{\Psi,P}(z)$. The following theorem is a direct consequence of Proposition 4.2 and Theorem 4.1. It characterizes the set of trees described by the graph $\Delta_{\Psi,P}(z)$.

Theorem. *Let U be an information system, Ψ a weight function, z a problem over U, and P a probability distribution for z. Then the algorithm $\mathcal{A}$ given* the extended table T_z^* builds a graph $\Delta_{\Psi,P}(z)$ that describes the set of all irredundant decision trees that are optimal relative to the average weighted depth.

For an arbitrary polynomial Q, a probability distribution P is called Q-restricted if for an arbitrary row $\bar{d} \in T_z$, the length of the binary notation of the number $P(\bar{d})$ does not exceed $Q(n)$ where n is the number of columns in the table. One more theorem formulated in Sect. 4.1 characterizes the time complexity of the algorithm $\mathcal{A}$.

Theorem. *(Theorem 4.2 from Sect. 4.1) Let $Q(x)$ be some polynomial. Then for an arbitrary problem $z = (\nu, f_1, \ldots, f_n)$ and an arbitrary Q-restricted probability distribution P for the problem z, the working time of the algorithm $\mathcal{A}$ is proportional to the number of rows $D(T_z)$ if the table T_z^* is terminal. If the table T_z^* is nonterminal, the working time of the algorithm $\mathcal{A}$ is bounded from below by the maximum of the values n, the number of nonterminal separable subtables $|S(z)|$, $D(T_z)$ and the maximum length of attribute weight in binary notation, and is bounded from above by a polynomial on these values.*

1.2.4   Restricted Information Systems

Chapter 5.2 among all other information systems distinguishes so-called restricted information systems. The property of being restricted implies a common upper bound on the minimum average weighted depth of decision tree that depends only on the entropy of probability distribution and holds for all problems over the information system. Another property of restricted information systems is that under reasonable assumptions about weight function and probability distribution, the working time of the algorithm $\mathcal{A}$ is limited from above by a polynomial on the number of attributes in the problem description.

For an arbitrary natural number t, a system of equations of the form $\{f_1(x) = \delta_1, \ldots, f_t(x) = \delta_t\}$ where $f_1, \ldots, f_t \in F$ and $\delta_1, \ldots, \delta_t \in E_k$, is called *a system of equations over U*. An information system U is called *r-restricted (restricted)* if each compatible system of equations over U has an equivalent subsystem that contains at most r equations.

For a system of equations $\{f_1(x) = \delta_1, \ldots, f_t(x) = \delta_t\}$ over the information system U, the value $\sum_{i=1}^{t} \Psi(f_i)$ is called *the weight of the system of equations*.

An information system U is called *r-restricted (restricted) relative to Ψ* if each compatible system of equations over U has an equivalent subsystem whose weight does not exceed r.

Example. (Example 5.1 from Sect. 5.1) Let $A = R^n$, and F be a nonempty set of mappings from R^n to R. Consider an infinite family of functions $[F] = \{\text{sign}(f + \alpha) + 1 : f \in F, \alpha \in R\}$ (note that the expression $(\text{sign}(x) + 1)$ takes the value 0 for a negative x, 1 for $x = 0$, and 2 for a positive x). If $|F| = k < \infty$, then the information system $U = (A, [F])$ is $2k$-restricted (or $2k$-restricted relative to the weight function $\Psi \equiv 1$).

The following theorem for an arbitrary problem over a restricted information system and an arbitrary probability distribution, gives an upper bound on the minimum average weighted depth of decision tree that depends only on the entropy of probability distribution.

Theorem. *(Theorem 5.1 from Sect. 5.1) Let U be an information system, Ψ a weight function for U, and U be r-restricted relative to Ψ where r is some natural number. Then $h_\Psi(z, P) \leq 2r(H(P) + 1)$ for an arbitrary problem z over U and an arbitrary probability distribution P for z.*

The following theorem shows that the conditions of Theorem 5.1 are necessary and sufficient for existence of a linear upper bound depending only on the entropy and considering non-linear bounds does not extend the class of information systems that have upper bounds depending only on the entropy.

Theorem. *(Theorem 5.2 from Sect. 5.1) Let U be an information system that is not restricted relative to the weight function Ψ for U. Then for an arbitrary $\varepsilon > 0$, there is no function Φ that is limited within the interval $[0, \varepsilon]$ and possesses the condition $h_\Psi(z, P) \leq \Phi(H(P))$ for any problem z over U and any probability distribution P for z.*

Denote $\mathcal{Z}(U)$ the set of problems over the information system U. For an arbitrary problem z, denote by $\dim z$ the number of attributes listed in the description of z.

Consider the functions

$$\mathcal{S}_U(n) = \max\{|S(z)| : z \in \mathcal{Z}(U), \dim z \leq n\}$$

and

$$\mathcal{D}_U(n) = \max\{D(T_z) : z \in \mathcal{Z}(U), \dim z \leq n\}$$

that characterize the dependence of the maximum number of separable subtables and the maximum number of rows on the number of columns in decision tables over U.

Let Ψ be restricted from above by some constant, and $Q(x)$ be some polynomial. Theorem 4.2 implies that for an arbitrary problem z over U and an arbitrary Q-restricted probability distribution for the problem z, the time complexity of the algorithm $\mathcal{A}$ is restricted from above by a polynomial on the number of attributes in the problem description if the functions $\mathcal{S}_U(n)$ and $\mathcal{D}_U(n)$ are restricted from above by a polynomial on n. Also, one can see that the time complexity of the algorithm $\mathcal{A}$ has an exponential lower bound if the function $\mathcal{S}_U(n)$ grows exponentially.

Theorem. *(Theorem 5.3 from Sect. 5.2) Let $U = (A, F)$ be a k-valued information system. Then the following statements hold:*

a) if U is r-restricted, then $\mathcal{S}_U(n) \leq (nk)^r + 1$ and $\mathcal{D}_U(n) \leq (nk)^r + 1$ for any natural number n;

b) if U is not restricted, then $\mathcal{S}_U(n) \geq 2^n - 1$ for any natural number n.

Chapter 2

Bounds on Average Time Complexity of Decision Trees

In this chapter, bounds on the average depth and the average weighted depth of decision trees are considered. Similar problems are studied in search theory [1], coding theory [77], design and analysis of algorithms (e.g., sorting) [38]. For any diagnostic problem, the minimum average depth of decision tree is bounded from below by the entropy of probability distribution (with a multiplier $1/\log_2 k$ for a problem over a k-valued information system). Among diagnostic problems, the problems with a complete set of attributes have the lowest minimum average depth of decision trees (e.g, the problem of building optimal prefix code [1] and a blood test study in assumption that exactly one patient is ill [23]). For such problems, the minimum average depth of decision tree exceeds the lower bound by at most one. The minimum average depth reaches the maximum on the problems in which each attribute is "indispensable" [44] (e.g., a diagnostic problem with n attributes and k^n pairwise different rows in the decision table and the problem of implementing the modulo 2 summation function). These problems have the minimum average depth of decision tree equal to the number of attributes in the problem description.

We also consider a possibility of problem decomposition. Some problems have a hierarchy of attributes: "basic" attributes perform a rough classification, and "extended" ones can be applied to refine the solution. In this case, the leaf composition [44] can be applied: a tree for rough classification is built using basic attributes only, and then each leaf is replaced with a tree that does fine classification using extended attributes only. We are interested in finding the conditions that make the tree resulted from such composition to have the minimum average time complexity. In this case, applying problem decomposition leads to both comprehensive and effective solution.

The chapter consists of four sections. The first section gives a known bound for a diagnostic problem with a complete set of attributes. The second section

I. Chikalov: Average Time Complexity of Decision Trees, ISRL 21, pp. 15–39.
springerlink.com   © Springer-Verlag Berlin Heidelberg 2011

generalizes the known lower bound and gives an upper bound for the average weighted depth, which depends on the parameter $M(z)$ and the entropy of probability distribution. The third section gives more precise upper bound for the minimum average depth of decision tree. The fourth section describes sufficient conditions for problem decomposition which allow synthesizing an optimal tree for the initial problem from optimal trees for subproblems. An example of decomposable problem is considered that has the minimum average depth of decision tree close to the upper bound given in Sect. 2.3. The results of this chapter were previously published in [16, 18, 51, 52, 53, 54].

2.1   Known Bounds

A problem $z = (\nu, f_1, \ldots, f_n)$ with s equivalence classes $Q_1, \ldots, Q_s$ over a k-valued information system $U = (A, F)$ contains a *complete set of attributes* if for an arbitrary partition $\{1, \ldots, s\} = \bigcup_{j \in E_k} I_j$ (where $I_i \cap I_j = \emptyset$ if $i \neq j$), there exists an attribute $f_t \in \{f_1, \ldots, f_n\}$ such that

$$\bigcup_{i \in I_j} Q_i = \{a \in A : f_t(a) = j\}$$

for each $j \in E_k$.

The following theorem gives a bound on the average depth of decision tree for a diagnostic problem that contains a complete set of attributes. The bound follows from coding theory results and is well known in search theory (see, for example, [1]).

Theorem 2.1. *Let z be a diagnostic problem with a complete set of attributes over a k-valued information system U, and P a probability distribution for z. Then*

$$\frac{H(P)}{\log_2 k} \leq h(z, P) \leq \frac{H(P)}{\log_2 k} + 1 .$$

2.2   Bounds on Average Weighted Depth

The following theorem generalizes the lower bound to the case of the average weighted depth of decision tree for an arbitrary diagnostic problem.

Theorem 2.2. *Let U be a k-valued information system, Ψ a weight function for U, z a diagnostic problem over U, and P a probability distribution for z. Then*

$$h_\Psi(z, P) \geq \frac{H(P)}{\log_2 k} .$$

Proof. Let $U = (A, F)$, $z = (\nu, f_1, \ldots, f_n)$, and the problem z contains s equivalence classes $Q_1, \ldots, Q_s$. Let $(I_0^1, \ldots, I_{k-1}^1), \ldots, (I_0^r, \ldots, I_{k-1}^r)$ be all possible partitions of the set $\{1, \ldots, s\}$ possessing the following conditions: $\cup_{j=0}^{k-1} I_j^t = \{1, \ldots, s\}$, and for any numbers $i, j \in E_k$, $i \neq j$, for $t = 1, \ldots, r$, the relation $I_i^t \cap I_j^t = \emptyset$ holds. Define an attribute $g_t : A \to E_k$, $t = 1, \ldots, r$, as follows. If $a \in Q_i$ and $i \in I_j^t$, then $g_t(a) = j$. Consider the problem $z' = (\nu', f_1, \ldots, f_n, g_1, \ldots, g_r)$ over the information system $U' = (A, F \cup \{g_1, \ldots, g_r\})$ where $\nu' : E_k^{n+r} \to \omega$, and $\nu'(\delta_1, \ldots, \delta_n, \delta_{n+1}, \ldots, \delta_{n+r}) = \nu(\delta_1, \ldots, \delta_n)$ for each $(\delta_1, \ldots, \delta_{n+r}) \in E_k^{n+r}$. According to the definition of the attributes $g_1, \ldots, g_r$, we have that the problem z' contains a complete set of attributes, and z' has the same equivalence classes as the problem z. Evidently, $z'(a) = z(a)$ for any element $a \in A$. Then z' is a diagnostic problem and Theorem 2.1 implies

$$h(z', P) \geq \frac{H(P)}{\log_2 k} . \tag{2.1}$$

Let Γ be a decision tree for the problem z that solves z. One can see that Γ is a decision tree for the problem z' that solves z'. Then

$$h(z, P) \geq h(z', P) . \tag{2.2}$$

Since $\Psi(f) \geq 1$ for an arbitrary attribute $f \in F$, the relation $h_\Psi(z, P) \geq h(z, P)$ holds. The last inequality, (2.1) and (2.2) imply the bound given by the theorem statement. $\qquad\square$

The following theorem gives an upper bound on the minimum average weighted depth of decision tree for an arbitrary problem.

Theorem 2.3. *Let U be an information system, Ψ a weight function for U, z a problem over U, and P a probability distribution for z. Then*

$$h_\Psi(z, P) \leq M_\Psi(z) H(P) + M_\Psi(z) .$$

The following proposition shows that the additive constant $M_\Psi(z)$ in the right part of the inequality is inherent.

Proposition 2.1. *For an arbitrary $m \in \omega \setminus \{0\}$, there exists a 2-valued information system U, a weight function Ψ for U, a problem z over U and a sequence of probability distributions $P_1, P_2, \ldots$ for z, such that $M_\Psi(z) = m$, $\lim_{i \to \infty} H(P_i) = 0$, and $\lim_{i \to \infty} h_\Psi(z, P_i) = m$.*

Proof. Let $m \in \omega \setminus \{0\}$. Define a 2-valued information system U as follows: $U = (A, F)$ where $A = \{0, 1, \ldots, m\}$, $F = \{f_1, \ldots, f_m\}$ and

$$f_i(a) = \begin{cases} 1, & \text{if } i = a, \\ 0, & \text{if } i \neq a, \end{cases}$$

for any $f_i \in F$ and $a \in A$. Assume that $\Psi(f_i) = 1$ for $i = 1, \ldots, m$. Let $z = (\nu, f_1, \ldots, f_m)$ be a diagnostic problem. One can see that z has $(m + 1)$ equivalence classes $Q_0 = \{0\}, Q_1 = \{1\}, \ldots, Q_m = \{m\}$, the table T_z contains $(m+1)$ rows and it is not a terminal table. Consider a probability distribution P_i for z, defined as follows:

$$P_i(\bar{\delta}) = \begin{cases} i, & \text{if } \bar{\delta} = (0, 0, \ldots, 0), \\ 1, & \text{if } \bar{\delta} \in T_z \setminus \{(0, 0, \ldots, 0)\}. \end{cases}$$

One can see that $\lim_{i \to \infty} H(P_i) = 0$. Let $\bar{\delta} \in E_2^m$. It is easy to show, that $M_\Psi(z, \bar{\delta}) = 1$ for $\bar{\delta} \neq (0, \ldots, 0)$, and $M_\Psi(z, \bar{\delta}) = m$ for $\bar{\delta} = (0, \ldots, 0)$. Consequently, $M_\Psi(z) = m$.

Let $i \in \omega \setminus \{0\}$, Γ be a decision tree for the problem z that solves z, and has $h_\Psi(\Gamma, P_i) = h_\Psi(z, P_i)$. Consider a complete path ξ in Γ such that $(0, \ldots, 0) \in T_z \pi(\xi)$. One can see that the length of the path ξ is at least m. Consequently, $h_\Psi(\Gamma, P_i) \geq mi/(i + m)$, and $h_\Psi(z, P_i) \geq mi/(i + m)$. Theorem 2.3 implies $h_\Psi(z, P_i) \leq M_\Psi(z)(H(P_i) + 1) = m(H(P_i) + 1)$. Using these relations, we have that $\lim_{i \to \infty} h_\Psi(z, P_i) = m$. □

2.3 Upper Bound on Average Depth

Since the average depth of decision tree is a particular case of the average weighted depth, the above considered upper and lower bounds hold for the average depth as well. However, the upper bound on the average depth can be improved.

Theorem 2.4. *Let z be a problem over an information system U, and P a probability distribution for z. Then*

$$h(z, P) \leq \begin{cases} M(z), & \text{if } M(z) \leq 1, \\ M(z) + 2H(P), & \text{if } 2 \leq M(z) \leq 3, \\ M(z) + \frac{M(z)}{\log_2 M(z)} H(P), & \text{if } M(z) \geq 4. \end{cases}$$

Theorem 2.6 in Sect. 2.4.3 characterizes quality of the obtained bound.

2.3.1   *Process of Building Decision Trees* $Y_{U,\Psi}$

Let $U = (A, F)$ be a k-valued information system, Ψ a weight function for U, $z = (\nu, f_1, \ldots, f_n)$ a problem over U, and P a probability distribution for z. In this section, a process $Y_{U,\Psi}$ is considered that takes as input z and P, and builds a decision tree $Y_{U,\Psi}(z, P)$ that solves the problem z. The bounds given by Theorem 2.3 and Theorem 2.4 are resulted from analysis of decision trees built by this process.

The set F can be uncountable and the function Ψ can be incomputable, so in general case, the process $Y_{U,\Psi}$ is a way of defining the decision tree $Y_{U,\Psi}(z, P)$ rather than an algorithm.

The process $Y_{U,\Psi}$ includes a subprocess X_Ψ that builds a decision tree $X_\Psi(z, P, T)$ by given z, P and an arbitrary nonterminal subtable T of the table T_z.

Define a mapping $\mathrm{num}_z : \Omega_z^* \to \omega$. For $j = 1, 2, \ldots$, denote by r_j the j-th prime number. Let $\beta \in \Omega_z^*$. If $\beta = \lambda$, then $\mathrm{num}_z(\beta) = 1$. Let $\beta \neq \lambda$ and $\beta = (f_{i_1}, \delta_1) \ldots (f_{i_t}, \delta_t)$. Then $\mathrm{num}_z(\beta) = r_1^{i_1} \times \ldots \times r_t^{i_t}$. The number $\mathrm{num}_z(\beta)$ will be called *z-number of the word* β.

For an arbitrary word $\alpha \in \Omega_z^*$, denote $h(\alpha)$ the length of the word α and denote $\chi(\alpha)$ the set of letters from the alphabet Ω_z that are contained in α.

Description of the subprocess X_Ψ

Let the subprocess X_Ψ be applied to the triplet z, P, T, where T is a nonterminal subtable of the table T_z.

Step 1. For each $i \in \{1, \ldots, n\}$, assume σ_i to be the minimum number σ from E_k for which

$$N(T(f_i, \sigma), P) = \max\{N(T(f_i, \delta), P) : \delta \in E_k\} \ .$$

Denote $\bar{\sigma} = (\sigma_1, \ldots, \sigma_n)$. Let β be the word with the minimum z-number among all words in Ω_z^* possessing the following conditions: $\chi(\beta) \subseteq \{(f_1, \sigma_1), \ldots, (f_n, \sigma_n)\}$, the subtable $T\beta$ is terminal, and $\Psi(\beta) = M_\Psi(z, \bar{\sigma})$. Note that $\beta \neq \lambda$, because the subtable T is nonterminal. Let $\beta = (f_{i_1}, \sigma_{i_1}) \ldots (f_{i_m}, \sigma_{i_m})$. Denote $I_1 = \{f_{i_1}, \ldots, f_{i_m}\}$. Build a tree that consists of a single node. Assign the word λ to this node. Denote G_1 the obtained tree. Proceed to the step 2.

Let $t \geq 1$ steps have been already done and a tree G_t and a set I_t have been built.

Step $(t+1)$. Find in the tree G_t the only node w that is assigned with a word from Ω_z^*. Denote α the word assigned to w.

If $I_t = \emptyset$, then assign to w the number 0 instead of the word α. Denote the resulted tree $X_\Psi(z, P, T)$. The subprocess X_Ψ is completed.

Let $I_t \neq \emptyset$. Let j be the minimum number form the set $\{1, \ldots, n\}$ possessing the following conditions: $f_j \in I_t$ and

$$\max\{N(T\alpha(f_j, \sigma), P) : \sigma \in E_k \setminus \{\sigma_j\}\}$$
$$\geq \max\{N(T\alpha(f_i, \sigma), P) : \sigma \in E_k \setminus \{\sigma_i\}\}$$

for any $f_i \in I_t$. Assign the attribute f_j to the node w instead of the word α. For each $\sigma \in E_k$, add to the tree G_t a node w_σ and the edge that leaves the node w and enters w_σ. Assign the number σ to that edge. Label the node w_σ with the word $\alpha(f_j, \sigma_j)$ if $\sigma = \sigma_j$, or with the number 0 otherwise. Denote by G_{t+1} the resulted tree. Assume $I_{t+1} = I_t \setminus \{f_j\}$. Proceed to the step $(t+2)$.

Description of the process $Y_{U,\psi}$

Let the process $Y_{U,\psi}$ be applied to the pair (z, P).

Step 1. Assume $T = T_z$. Build a decision tree that consists of a single node v.

Let T be a terminal table. Then assign the number $\nu(\bar{\delta})$ to the node v where $\bar{\delta}$ is an arbitrary row from T. Denote $Y_{U,\psi}(z, P)$ the resulted decision tree. The process $Y_{U,\psi}$ is completed.

Let T be a nonterminal table. Assign the word λ to the node v and proceed to the next step.

Let $t \geq 1$ steps have been already done. Denote G the tree built at the step t.

Step $(t+1)$. If no node in G is assigned with a word from Ω_z^*, then denote $Y_{U,\psi}(z, P)$ the tree G. The process $Y_{U,\psi}$ is completed. Otherwise, choose in G a terminal node v, which is assigned with a word from Ω_z^*. Denote α the word assigned to v.

Let $T\alpha$ be a terminal subtable. If $T\alpha = \emptyset$, then assign to v the number 0 instead of the word α. If $T\alpha \neq \emptyset$, then assign to v the number $\nu(\bar{\delta})$ instead of α where $\bar{\delta}$ is an arbitrary row from $T\alpha$. Proceed to the step $(t+2)$.

Let $T\alpha$ be a nonterminal subtable. Apply the subprocess X_ψ to build the decision tree $X_\psi(z, P, T\alpha)$. For each complete path ξ in $X_\psi(z, P, T\alpha)$, replace the number 0 assigned to its terminal node, with the word $\alpha\pi(\xi)$. Denote Γ the tree resulted from this replacement. Replace in G the node v with the tree Γ. Proceed to the step $(t+2)$.

2.3.2   *Proofs of Theorems 2.3 and 2.4*

This section contains proofs of the upper bounds on the minimum average time complexity of decision trees given in Sect. 2.2 and Sect. 2.3.

Lemma 2.1. *Let $U = (A, F)$ be a k-valued information system, Ψ a weight function for U, $z = (\nu, f_1, \ldots, f_n)$ a problem over U, P a probability distribution for z, and T a nonterminal subtable of the table T_z. Then the following conditions hold for each complete path ξ in the decision tree $X_\Psi(z, P, T)$:*

a) $\Psi(\pi(\xi)) \leq M_\Psi(z)$;

b) if $T\pi(\xi)$ is a nonterminal subtable, then

$$N(T\pi(\xi), P) \leq N(T, P) / \max\{2, h(\pi(\xi))\} .$$

Proof. For each $i \in \{1, \ldots, n\}$, denote σ_i the minimum number from E_k such that

$$N(T(f_i, \sigma_i), P) = \max\{N(T(f_i, \sigma), P) : \sigma \in E_k\} .$$

Denote $\bar{\sigma} = (\sigma_1, \ldots, \sigma_n)$. Denote β a word from Ω_z^* with the minimum z-number possessing the following conditions: $\chi(\beta) \subseteq \{(f_1, \sigma_1), \ldots, (f_n, \sigma_n)\}$, $T\beta$ is a terminal table, and $\Psi(\beta) = M_\Psi(z, \bar{\sigma})$. Obviously, all letters in the word β are pairwise different. Using this property of the word β and the description of the subprocess X_Ψ, one can show that there exists a complete path ξ_0 in the tree $X_\Psi(z, P, T)$ such that $\chi(\pi(\xi_0)) = \chi(\beta)$, and the words $\pi(\xi_0)$ and β are of equal length. Then $T\pi(\xi_0)$ is a terminal subtable, and $\Psi(\pi(\xi_0)) = \Psi(\beta)$. Taking into account the choice of the word β, we have

$$\Psi(\pi(\xi_0)) = M_\Psi(z, \bar{\sigma}) . \tag{2.3}$$

Let $\pi(\xi_0) = (f_{j_1}, \sigma_{j_1}) \ldots (f_{j_m}, \sigma_{j_m})$. Denote $\alpha_0 = \lambda$, and for $i = 1, \ldots, m$, denote $\alpha_i = (f_{j_1}, \sigma_{j_1}) \ldots (f_{j_i}, \sigma_{j_i})$. For $i = 1, \ldots, m$, denote δ_{j_i} the minimum number from $E_k \setminus \{\sigma_{j_i}\}$ such that

$$N(T\alpha_{i-1}(f_{j_i}, \delta_{j_i}), P) = \max\{N(T\alpha_{i-1}(f_{j_i}, \sigma), P) : \sigma \in E_k \setminus \{\sigma_{j_i}\}\} .$$

Let ξ be an arbitrary complete path in the decision tree $X_\Psi(z, P, T)$. Let $\xi = \xi_0$. By applying (2.3), we obtain $\Psi(\pi(\xi_0)) = M_\Psi(z, \bar{\sigma}) \leq M_\Psi(z)$. Let $\xi \neq \xi_0$. One can see that there exist numbers $r \in \{1, \ldots, m\}$ and $\delta \in E_k$ such that $\pi(\xi) = \alpha_{r-1}(f_{j_r}, \delta)$. Therefore, $\Psi(\pi(\xi)) \leq \Psi(\pi(\xi_0))$ and $\Psi(\pi(\xi)) \leq M_\Psi(z)$. Part (a) of the lemma is proved.

Let ξ be an arbitrary complete path in the decision tree $X_\Psi(z, P, T)$, for which $T\pi(\xi)$ is a nonterminal subtable. The fact that the subtable $T\pi(\xi_0)$ is terminal implies $\xi \neq \xi_0$. It is easy to see that there exist numbers $r \in \{1, \ldots, m\}$ and $\delta \in E_k$ such that $\delta \neq \sigma_{j_r}$ and $\pi(\xi) = \alpha_{r-1}(f_{j_r}, \delta)$.

Let us show that $N(T\pi(\xi), P) \leq N(T, P)/2$. Evidently,

$$N(T\pi(\xi), P) \leq N(T(f_{j_r}, \delta), P) .$$

Taking into account the choice of the word σ_{j_r}, we obtain that $N(T(f_{j_r}, \delta),$ $P) \leq N(T(f_{j_r}, \sigma_{j_r}), P)$. Since $\delta \neq \sigma_{j_r}$, the relation

$$N(T(f_{j_r}, \delta), P) + N(T(f_{j_r}, \sigma_{j_r}), P) \leq N(T, P)$$

holds. Consequently, $N(T\pi(\xi), P) \leq N(T, P)/2$.

Obviously, $h(\pi(\xi)) = r$. Let $r \geq 2$. Let us show that $N(T\pi(\xi), P) \leq N(T, P)/r$. Since $\delta_{j_{i+1}} \neq \sigma_{j_{i+1}}$, the inequalities

$$N(T\alpha_{i+1}, P) + N(T\alpha_i(f_{j_{i+1}}, \delta_{j_{i+1}}), P) \leq N(T\alpha_i, P)$$

hold for $i = 0, \ldots, r - 2$. Summing these inequalities by i from 0 to $r - 2$, we obtain

$$N(T\alpha_{r-1}, P) + \sum_{i=0}^{r-2} N(T\alpha_i(f_{j_{i+1}}, \delta_{j_{i+1}}), P) \leq N(T, P) . \tag{2.4}$$

Let us show that for any $i \in \{0, \ldots, r - 2\}$,

$$N(T\pi(\xi), P) \leq N(T\alpha_i(f_{j_{i+1}}, \delta_{j_{i+1}}), P) . \tag{2.5}$$

The inequality

$$N(T\alpha_i(f_{j_r}, \delta), P) \leq N(T\alpha_i(f_{j_{i+1}}, \delta_{j_{i+1}}), P)$$

follows from the choice of the attribute $f_{j_{i+1}}$ (see description of the subprocess X_Ψ) and the definition of the number $\delta_{j_{i+1}}$. The inequality

$$N(T\pi(\xi), P) \leq N(T\alpha_i(f_{j_r}, \delta), P)$$

is obvious. These two inequalities imply (2.5). The inequality $N(T\pi(\xi), P) \leq N(T\alpha_{r-1}, P)$ is obvious. This inequality, (2.4) and (2.5) imply $rN(T\pi(\xi), P) \leq N(T, P)$. Since $r \geq 2$, the relation $N(T\pi(\xi), P) \leq N(T, P)/r$ holds. Part (b) of the lemma is proved. □

Using the description of the process $Y_{U,\Psi}$ and subprocess X_Ψ, and Lemma 2.1, it is not hard to prove the following proposition.

Proposition 2.2. *Let U be an information system, and Ψ a weight function for U. Then for any problem z over U and any probability distribution P for z, the process $Y_{U,\Psi}$ ends in a finite number of steps. The resulted tree $Y_{U,\Psi}(z, P)$ is a decision tree for the problem z that solves z.*

Proof of Theorem 2.3. If T_z is a terminal table, then the equality $h_\Psi(Y_{U,\Psi}(z, P)) = 0$ follows from the description of the process $Y_{U,\Psi}$. This equality and Proposition 2.2 imply $h_\Psi(z, P) \leq 0$.

Let T_z be a nonterminal table. Consider an arbitrary row $\bar{d} \in T_z$ and find the complete path $\xi^{\bar{d}}$ in the decision tree $Y_{U,\Psi}(z, P)$ such that $\bar{d} \in T_z\pi(\xi^{\bar{d}})$. From the description of the process $Y_{U,\Psi}$ and the assumption that T_z is a nonterminal table it follows that $\pi(\xi^{\bar{d}}) = \pi(\xi_1^{\bar{d}})\ldots\pi(\xi_m^{\bar{d}})$ for some $m \in \omega\setminus\{0\}$ where $\xi_1^{\bar{d}}$ is a complete path in the decision tree $X_\Psi(z, P, T)$, and (if $m \geq 2$) $\xi_i^{\bar{d}}$ is a complete path in the decision tree $X_\Psi(z, P, T\pi(\xi_1^{\bar{d}})\ldots\pi(\xi_{i-1}^{\bar{d}}))$, $i = 2, \ldots, m$.

By the assumption, the table T_z is nonterminal. If $m \geq 2$, then the description of the process $Y_{U,\Psi}$ implies $T\pi(\xi_1^{\bar{d}})\ldots\pi(\xi_{i-1}^{\bar{d}})$ is a nonterminal table for $i = 2, \ldots, m$. Applying part (a) of Lemma 2.1, we obtain $\Psi(\pi(\xi_i^{\bar{d}})) \leq M_\Psi(z)$ for $i = 1, \ldots, m$. Consequently,

$$\Psi(\pi(\xi^{\bar{d}})) = \sum_{i=1}^{m} \Psi(\pi(\xi_i^{\bar{d}})) \leq mM_\Psi(z) . \tag{2.6}$$

Let us show that $m \leq -\log_2 P(\bar{d}) + \log_2 N(T_z, P) + 1$. Evidently, the inequality holds for $m = 1$. Let $m \geq 2$. Part (b) of Lemma 2.1 implies

$$N(T_z\pi(\xi_1^{\bar{d}})\ldots\pi(\xi_{m-1}^{\bar{d}}), P) \leq \frac{N(T_z, P)}{2^{m-1}} .$$

One can see that $\bar{d} \in T_z\pi(\xi_1^{\bar{d}})\ldots\pi(\xi_{m-1}^{\bar{d}})$. Taking into account this condition, we obtain

$$N(T_z\pi(\xi_1^{\bar{d}})\ldots\pi(\xi_{m-1}^{\bar{d}}), P) \geq P(\bar{d}) .$$

Consequently, $2^{m-1} \leq N(T_z, P)/P(\bar{d})$ and $m \leq -\log_2 P(\bar{d}) + \log_2 N(T_z, P) + 1$. The obtained inequality and (2.6) result in

$$\Psi(\pi(\xi^{\bar{d}})) \leq M_\Psi(z)(-\log_2 P(\bar{d}) + \log_2 N(T_z, P) + 1) .$$

From the definition of the weighted average depth it follows that

$$h_\Psi(Y_{U,\Psi}(z, P), P) = \frac{1}{N(T_z, P)} \sum_{\bar{d}\in T_z} \Psi(\pi(\xi^{\bar{d}}))P(\bar{d})$$

$$\leq \frac{1}{N(T_z, P)} M_\Psi(z) \sum_{\bar{d}\in T_z} (\log_2 N(T_z, P) - \log_2 P(\bar{d}) + 1)P(\bar{d})$$

$$= M_\Psi(z)(H(P) + 1) .$$

This inequality and Proposition 2.2 imply $h_\Psi(z, P) \leq M_\Psi(z)(H(P) + 1)$. $\square$

Proof of Theorem 2.4. Let $z = (\nu, f_1, \ldots f_n)$. If $M(z) = 0$, one can see that the table T_z is terminal and $h(z, P) = 0$.

Let $M(z) = 1$. Assume that for $i = 1, \ldots, n$, there exists a number $\delta_i \in E_k$ such that $\nu(x) \not\equiv$ const on the set of rows of the subtable $T_z(f_i, \delta_i)$. Denote $\bar{\delta} = (\delta_1, \ldots, \delta_n)$. One can see that $M(z, \bar{\delta}) \geq 2$, but, according to the definition, $M(z, \bar{\delta}) \leq M(z) = 1$. This contradiction shows that there exists an attribute $f_i \in \{f_1, \ldots, f_n\}$ such that for any $\delta \in E_k$, either $T_z(f_i, \delta)$ is empty or $\nu(x) \equiv$ const on the set of rows of this table. It is easy to show that there exists a decision tree Γ for the problem z that solves z for which $h(\Gamma, P) = 1$. Consequently, $h(z, P) \leq 1$.

Let $M(z) \geq 2$. This inequality requires T_z to be a nonterminal table. Consider an arbitrary row $\bar{d} \in T_z$ and find a complete path $\xi^{\bar{d}}$ in the decision tree $Y_{U,h}(z, P)$ such that $\bar{d} \in T_z\pi(\xi^{\bar{d}})$. From the description of the process $Y_{U,h}$ and from the fact that T_z is a nonterminal subtable it follows that $\pi(\xi^{\bar{d}}) = \pi(\xi_1^{\bar{d}}) \ldots \pi(\xi_m^{\bar{d}})$ for some $m \in \omega \setminus \{0\}$ where $\xi_1^{\bar{d}}$ is a complete path in the decision tree $X_h(z, P, T_z)$, and (if $m \geq 2$) $\xi_i^{\bar{d}}$ is a complete path in the decision tree $X_h(z, P, T\pi(\xi_1^t) \ldots \pi(\xi_{i-1}^t))$, $i = 2, \ldots, m$. Denote $r_i^{\bar{d}} = h(\pi(\xi_i^{\bar{d}}))$ for $i = 1, \ldots, m$. Let us estimate the value $h(\pi(\xi^{\bar{d}})) = \sum_{i=1}^m r_i^{\bar{d}}$. We will prove that

$$h(\pi(\xi^{\bar{d}})) \leq \begin{cases} -2\log_2 P(\bar{d}) + 2\log_2 N(T_z, P) \\ \qquad\qquad + M(z), & \text{if } 2 \leq M(z) \leq 3, \\[2ex] \dfrac{M(z)}{\log_2 M(z)}(-\log_2 P(\bar{d}) \\ + \log_2 N(T_z, P)) + M(z), & \text{if } M(z) \geq 4. \end{cases} \qquad (2.7)$$

Let $m = 1$. Part (a) of Lemma 2.1 implies that $r_1^{\bar{d}} \leq M(z)$. Therefore, the inequality (2.7) holds for $m = 1$. Let $m \geq 2$. Denote $y_i^{\bar{d}} = \max\{2, r_i^{\bar{d}}\}$ for $i = 1, \ldots, m$. By the assumption, T_z is a nonterminal table. From the description of the process $Y_{U,h}$ it follows that $T_z\pi(\xi_1^{\bar{d}}) \ldots \pi(\xi_i^{\bar{d}})$ is a nonterminal subtable for $i = 1, \ldots, m - 1$. Lemma 2.1 and inequality $m \geq 2$ imply

$$N(T_z\pi(\xi_1^{\bar{d}}) \ldots \pi(\xi_{m-1}^{\bar{d}}), P) \leq \frac{N(T_z, P)}{\prod_{i=1}^{m-1} y_i^{\bar{d}}}.$$

Since $\bar{d} \in T_z\pi(\xi_1^{\bar{d}}) \ldots \pi(\xi_{m-1}^{\bar{d}})$, we obtain $N(T\pi(\xi_1^{\bar{d}}) \ldots \pi(\xi_{m-1}^{\bar{d}}), P) \geq P(\bar{d})$. Consequently,

$$\prod_{i=1}^{m-1} y_i^{\bar{d}} \leq \frac{N(T_z, P)}{P(\bar{d})}.$$

Taking the binary logarithm of both sides results in

$$\sum_{i=1}^{m-1} \log_2 y_i^{\bar{d}} \le -\log_2 P(\bar{d}) + \log_2 N(T_z, P) \,.$$

This inequality implies

$$\sum_{i=1}^{m} r_i^{\bar{d}} = r_m^{\bar{d}} + \sum_{i=1}^{m-1} (\log_2 y_i^{\bar{d}} (r_i^{\bar{d}} / \log_2 y_i^{\bar{d}}))$$

$$\le r_m^{\bar{d}} + (\sum_{i=1}^{m-1} \log_2 y_i^{\bar{d}})(\max\{r_i^{\bar{d}} / \log_2 y_i^{\bar{d}} : i \in \{1, \dots, m-1\}\}) \qquad (2.8)$$

$$\le r_m^{\bar{d}} - (\log_2 P(d) - \log_2 N(T_z, P))$$

$$\times (\max\{r_i^{\bar{d}} / \log_2 y_i^{\bar{d}} : i \in \{1, \dots, m-1\}\}) \,.$$

Consider the function $q(x) = x / \log_2(\max\{2, x\})$, $x \in \omega \setminus \{0\}$. One can see that $q(0) = 0$, $q(1) = 1$, $q(2) = 2$, $q(3) < 2$, $q(4) = 2$, and the function $q(x)$ is monotonically increasing for $x \ge 3$. Therefore, for any $n \in \omega \setminus \{0\}$, the following condition holds:

$$\max\{q(i) : i \in \{0, \dots, n\}\} = \begin{cases} 1 \,, & \text{if } n = 1 \,, \\ 2 \,, & \text{if } 2 \le n \le 3 \,, \\ \frac{n}{\log_2 n} \,, & \text{if } n \ge 4 \,. \end{cases} \qquad (2.9)$$

From part (a) of Lemma 2.1 it follows that the inequality

$$r_i^{\bar{d}} \le M(z) \qquad (2.10)$$

holds for $i = 1, \dots, m$. From (2.9), (2.10) and the inequality $M(z) \ge 2$ we have

$$\max\{r_i^{\bar{d}} / \log_2 y_i^{\bar{d}} : i \in \{1, \dots, m-1\}\}$$

$$\le \max\{q(i) : i \in \{0, \dots, M(z)\}\} = \begin{cases} 2 \,, & \text{if } 2 \le M(z) \le 3 \,, \\ \frac{M(z)}{\log_2 M(z)} \,, & \text{if } M(z) \ge 4 \,. \end{cases}$$

These inequalities and inequalities (2.8) and (2.10) imply (2.7). Then

$$h(Y_{U,h}(z,P),P) = \sum_{\bar{d} \in T_z} h(\pi(\xi^{\bar{d}}))P(\bar{d})$$

$$\leq \begin{cases} M(z) + 2H(P) , & \text{if } 2 \leq M(z) \leq 3 , \\ M(z) + \frac{M(z)}{\log_2 M(z)}H(P) , & \text{if } M(z) \geq 4 . \end{cases}$$

Proposition 2.2 results in correctness of the theorem for $M(z) \geq 2$.   □

2.4   On Possibility of Problem Decomposition

In this section, a possibility of reduction is considered for a problem over 2-valued information system. Under certain conditions an optimal (relative to the average depth) decision tree can be constructed as a composition of optimal decision trees for simpler problems that form so-called proper decomposition of the original problem.

2.4.1   Proper Problem Decomposition

Let $U = (A, F)$ be a 2-valued information system, $z_0 = (\nu_0, f_1^0, \dots, f_{n_0}^0)$ a diagnostic problem over U with m classes of equivalence $A_1, \dots, A_m$. Let $T_{z_0} = \{\bar{d}_1^0, \dots, \bar{d}_m^0\}$ where $\bar{d}_i^0 = (f_1^0(a_i), \dots, f_{n_0}^0(a_i))$, $a_i \in A_i$, $i = 1, \dots, m$. For $i = 1, \dots, m$, let $z_i = (\nu_i, f_1^i, \dots, f_{n_i}^i)$ be a problem over the information system (A_i, F) with s_i classes of equivalence, and the table T_{z_i} contains s_i rows $\bar{d}_1^i, \dots, \bar{d}_{s_i}^i$. For $i = 1, \dots, m$, let P_i be an arbitrary probability distribution for the problem z_i, and $P_0 = (N(T_{z_1}, P_1), \dots, N(T_{z_m}, P_m))$ be a probability distribution for the problem z_0.

For $i = 0, \dots, m$ and $j = 1, \dots, s_i$, denote $\bar{\sigma}_j^i = (\bar{\alpha}_0^{i,j}\bar{\alpha}_1^{i,j} \dots \bar{\alpha}_m^{i,j})$ where $\bar{\alpha}_k^{i,j} \in E_2^{n_k}$, $k = 0, \dots, m$, and

$$\bar{\alpha}_k^{i,j} = \begin{cases} \bar{d}_i^0 , & \text{if } k = 0 , \\ \bar{d}_j^i , & \text{if } k = i , \\ (0, \dots, 0) , & \text{if } k \in \{1, \dots m\} \setminus \{i\} . \end{cases}$$

Define a function $\nu : E_2^{n_0 + \dots + n_m} \to \omega$ as follows:

$$\nu(\bar{\delta}) = \begin{cases} \nu_i(\bar{d}_j^i) , & \text{if } \bar{\delta} = \bar{\sigma}_j^i \text{ for some } i \in \{1, \dots, m\} \text{ and } j \in \{1, \dots, s_i\} , \\ 0 , & \text{otherwise} . \end{cases}$$

Consider a problem $z = (\nu, f_1^0, \ldots, f_{n_0}^0, \tilde{f}_1^1, \ldots, \tilde{f}_{n_1}^1, \ldots, \tilde{f}_1^m, \ldots, \tilde{f}_{n_m}^m)$ over U, where

$$\tilde{f}_j^i(a) = \begin{cases} f_j^i(a), & \text{if } a \in A_i, \\ 0, & \text{if } a \notin A_i, \end{cases}$$

for $j = 1, \ldots, n_i$, $i = 1, \ldots, m$ and $a \in A$. One can see that the table T_z contains the rows $\bar{\sigma}_1^1, \ldots, \bar{\sigma}_{s_1}^1, \ldots, \bar{\sigma}_1^m, \ldots, \bar{\sigma}_{s_m}^m$ and does not contain any other rows. Define a probability distribution P for the problem z as follows: $P(\bar{\sigma}_j^i) = P_i(\bar{d}_j^i)$ for $j = 1, \ldots, s_i$ and $i = 1, \ldots, m$. The set $((z_0, P_0), (z_1, P_1), \ldots, (z_m, P_m))$ is called *a proper decomposition of the pair* (z, P) if:

i) for $j = 1, \ldots, n_i$ and $i = 1, \ldots, m$, the inequality

$$N(T_{z_i}(f_j^i, 1), P_i) \leq \frac{\min_{l \in 1, \ldots, m} N(T_{z_l}, P_l)}{2}$$

   holds;

ii) for any $i, j \in \{1, \ldots, m\}$, $i \neq j$ and $c \in \omega$ such that

$$q_i = \sum_{\bar{d} \in T_{z_i}, \nu_i(\bar{d}) = c} P_i(\bar{d}) / N(T_{z_i}, P_i) > 0$$

   and

$$q_j = \sum_{\bar{d} \in T_{z_j}, \nu_j(\bar{d}) = c} P_j(\bar{d}) / N(T_{z_j}, P_j) > 0,$$

   the inequalities $\min(q_i, q_j) < 1/2$ and $\max(q_i, q_j) < 1$ hold.

Let $((z_0, P_0), (z_1, P_1), \ldots, (z_m, P_m))$ be a proper decomposition of the pair (z, P), and Γ_i be a decision tree for the problem z_i that solves z_i, $i = 0, \ldots, m$. For $i = 1, \ldots, m$, apply the following transformation to the tree Γ_i. For each nonterminal node w, replace the attribute f_j^i that is assigned to w with the corresponding attribute $\tilde{f}_j^i$. Denote the resulted tree by $\tilde{\Gamma}_i$.

For $i = 1, \ldots, m$, let us find a complete path ξ_i in Γ_0 such that $\bar{d}_i^0 \in T_{z_0}\pi(\xi_i)$ and replace the terminal node of the path ξ_i with the tree $\tilde{\Gamma}_i$. Denote the resulted tree by $\Phi(\Gamma_0, \Gamma_1, \ldots, \Gamma_m)$.

2.4.2   Theorem of Decomposition

Theorem 2.5. *Let z be a problem over a 2-valued information system U, P a probability distribution for z, and $((z_0, P_0), (z_1, P_1), \ldots, (z_m, P_m))$ a proper decomposition of the pair (z, P). Let Γ_i be a decision tree for z_i that solves z_i*

and is optimal for z_i and P_i, $i = 0, \ldots, m$. Then the tree $\Phi(\Gamma_0, \Gamma_1, \ldots, \Gamma_m)$ is a decision tree for the problem z that solves z and is optimal for z and P.

We preface proof of the theorem by a series of lemmas. Let us define some auxiliary notions.

Let Γ be a decision tree for the problem z. Denote $V(\Gamma)$ and $E(\Gamma)$ the set of nodes and the set of edges of Γ respectively. For an arbitrary node $v \in V(\Gamma)$, denote by $path(\Gamma, v)$ the path from the root of Γ to the node v. For an arbitrary nonterminal node $v \in V(\Gamma)$ and an arbitrary number $\delta \in \{0, 1\}$, denote by $e(\Gamma, v, \delta)$ the edge that leaves v and is labeled with δ. Let v be a nonterminal node in the tree Γ and f the attribute assigned to v. The node v is called *essential* if the table $T_z \pi(path(\Gamma, v))(f, \delta)$ is nonempty for $\delta = 0, 1$. The decision tree is called *reduced* if all its nonterminal nodes are essential.

Let z be a problem over information system, P a probability distribution for z, and $((z_0, P_0), (z_1, P_1), \ldots, (z_m, P_m))$ a proper decomposition for the pair (z, P). An attribute from the description of the problem z is called *basic* if it is contained in the description of z_0, or *extended* otherwise.

Let Γ be a reduced decision tree for the problem z, and $\xi = v_1, e_1, \ldots, v_t, e_t, v_{t+1}$ a complete path in Γ where $v_1, \ldots, v_{t+1} \in V(\Gamma)$ and $e_1, \ldots, e_t \in E(\Gamma)$, $t \geq 1$. Let for some $k \in \{1, \ldots, t+1\}$, the node v_i be assigned with a basic attribute if and only if $i < k$. Then the path ξ is called *ordered by basic attributes*.

Let the path ξ be not ordered by basic attributes. For $i = 1, \ldots, t$, denote f_i the attribute assigned to the node v_i. Then there exist natural j and k, $j < k \leq t$ such that $f_j, \ldots, f_{k-1}$ are extended attributes, f_k is a basic attribute, and (if $j > 1$) $f_1, \ldots, f_{j-1}$ are basic attributes. Denote

$$N_0(\xi) = N(T_z \pi(path(\Gamma, v_k))(f_k, 0), P)$$

and

$$N_1(\xi) = N(T_z \pi(path(\Gamma, v_k))(f_k, 1), P) \, .$$

Since v_k is assigned with a basic attribute and Γ is a reduced tree, we have that for $i = j, \ldots, k - 1$, the edge e_i is assigned with the number 0. For $i = j, \ldots, k - 1$, denote w_i the node that the edge $e(\Gamma, v_i, 1)$ enters, and denote σ_i the number such that $T_z \pi(path(\Gamma, w_i)) = T_z \pi(path(\Gamma, w_i))(f_k, \sigma_i)$. The path ξ is called *reducible* if $k \geq j+3$ and the set $\{\sigma_j, \ldots, \sigma_{k-1}\}$ contains both 0 and 1.

Define a *path reduction operation*. Let the operation be applied to the path ξ.

Step 1. For $\delta = 0, 1$, denote $\Gamma(\delta)$ the subtree that the edge $e(\Gamma, v_k, \delta)$ enters, and denote $e(\delta) = e(\Gamma, v_k, \delta)$. If v_j is not the root of Γ, then reroute the edge e_{j-1} so that it enters the node v_k. Proceed to the step 2.

Let i steps have already been done for some $1 \le i \le k - j$.

Step $(i + 1)$. Reroute the edge $e(\sigma_{j+i-1})$ so that it enters the node v_{j+i-1}. Denote $e(\sigma_{j+i-1}) = e(\Gamma, v_{j+i-1}, 0)$. Proceed to the step $(i + 2)$.

Step $(k - j + 2)$. For $\delta = 0, 1$, reroute the edge $e(\delta)$ so that it enters the subtree $\Gamma(\delta)$. The transformation is completed.

Lemma 2.2. *Let z be a problem over an information system, P a probability distribution for z and D a proper decomposition for the pair (z, P). Let Γ be a reduced decision tree for z that solves z, ξ a complete path in Γ that is not ordered by basic attributes, and $\tilde{\Gamma}$ the tree resulted from applying the path reduction operation to ξ. Then*

a) $\tilde{\Gamma}$ is a decision tree for z that solves z;
b) $h(\tilde{\Gamma}, P) \le h(\Gamma, P)$;
c) $h(\tilde{\Gamma}, P) \le h(\Gamma, P) - (N_0(\xi) + N_1(\xi))/N(T_z, P)$ if ξ is a reducible path.

Proof. One can see from the description of the path reduction operation that $\tilde{\Gamma}$ is a decision tree for the problem z. Let us show that $\tilde{\Gamma}$ solves z. Let $\xi = v_1, e_1, \ldots, v_t, e_t, v_{t+1}$ where $v_1, \ldots, v_{t+1} \in V(\Gamma)$, $e_1, \ldots, e_t \in E(\Gamma)$, $t \ge 1$, and for $i = 1, \ldots, t$, the node v_i is assigned with an attribute f_i. Then there exist natural j and k, $j < k \le t$, such that $f_j, \ldots, f_{k-1}$ are extended attributes, f_k is a basic attribute, and (if $j > 1$) $f_1, \ldots, f_{j-1}$ are basic attributes. Let $\bar{d} \in T_z$ be an arbitrary row, and ϕ a complete path in Γ such that $\bar{d} \in T_z\pi(\phi)$. Denote $\tilde{\phi}$ the complete path in $\tilde{\Gamma}$ that ends in the same terminal node as ϕ. Let us show that $\bar{d} \in T_z\pi(\tilde{\phi})$. If $v_j \notin \phi$, then $\tilde{\phi} = \phi$. If $v_k \in \phi$, then the path $\tilde{\phi}$ is obtained from ϕ by deleting several pairs consisting of a node and one of its outgoing edges. Therefore $T_z\pi(\tilde{\phi}) \supseteq T_z\pi(\phi)$. Let $v_i \in \phi$ and $v_{i+1} \notin \phi$ for some $i \in \{j, \ldots, k - 1\}$. Since $v_{i+1} \notin \phi$, the edge that leaves the node v_i and is contained in the path ϕ, is labeled with 1. The fact that f_i is an extended attribute implies that the set of solutions of the equation $f_i(x) = 1$ is contained in one of the equivalence classes of the problem z_0. Then there exists a number $\delta \in \{0, 1\}$ such that $T_z\pi(\phi) = T_z\pi(\phi)(f_k, \delta)$. From the description of the path reduction operation it follows that the path $\tilde{\phi}$ is obtained from ϕ by deleting several pairs consisting of a node and one of its outgoing edges and by adding the pair $(v_k, e(\Gamma, v_k, \delta))$. Therefore, $T_z\pi(\tilde{\phi}) \supseteq T_z\pi(\phi)(f_k, \delta)$, and, taking into account the last relation, $T_z\pi(\tilde{\phi}) \supseteq T_z\pi(\phi)$. In general case, $T_z\pi(\tilde{\phi}) \supseteq T_z\pi(\phi)$ and $\bar{d} \in T_z\pi(\tilde{\phi})$. Then the fact that Γ solves the problem z implies that $\tilde{\Gamma}$ also solves z.

Let us prove part (b) and (c) of the lemma. Since Γ is a reduced decision tree, v_k is an essential node. Then for $\delta = 0, 1$, the table $T_{z_0}\pi(path(\Gamma, v_j))$ contains at least one row in which the attribute f_k takes the value δ. Denote this row $\bar{d}^0_{i_\delta}$ and denote $P_\delta = N(T_{z_{i_\delta}}, P_{i_\delta})$. For $i = j, \ldots, k-1$, denote w_i the node which the edge $e(\Gamma, v_i, 1)$ enters, denote σ_i the number from the set $\{0, 1\}$ such that $T_z\pi(path(\Gamma, w_i)) = T_z\pi(path(\Gamma, w_i))(f_k, \sigma_i)$, and denote

$$N_i = N(T_z\pi(path(\Gamma, v_i)(f_i, 1)), P) .$$

Then for $\delta = 0, 1$, the following relation holds:

$$\sum_{i=j, \sigma_i = \delta}^{k-1} N_i + N_\delta(\xi) \geq P_\delta . \tag{2.11}$$

Consider several cases.

1) Let $\sigma_j = \ldots = \sigma_{k-1}$. One can see that ξ is not a reducible path. Then

$$h(\tilde{\Gamma}, P) = h(\Gamma, P) + \frac{1}{N(T_z, P)} \left(\sum_{i=j}^{k-1} N_i - (k-j)N_{1-\sigma_j}(\xi) \right) . \tag{2.12}$$

The relation (i) from the definition of the proper problem decomposition implies $N_i \leq P_{1-\sigma_j}$ for $i = j, \ldots, k-1$. Summing these inequalities, we obtain

$$\sum_{i=j}^{k-1} N_i \leq (k-j)P_{1-\sigma_j} . \tag{2.13}$$

From (2.11) it follows that

$$N_{1-\sigma_j}(\xi) \geq P_{1-\sigma_j} . \tag{2.14}$$

The relations (2.12), (2.13), (2.14) imply (b).

2) Let $k = j+2$ and $\sigma_j \neq \sigma_{j+1}$. One can see that ξ is not a reducible path. Then

$$h(\tilde{\Gamma}, P) = h(\Gamma, P) + \frac{N_j - N_{\sigma_{j+1}}(\xi)}{N(T_z, P)} . \tag{2.15}$$

The relation (i) implies

$$N_j + N_{j+1} \leq P_{\sigma_{j+1}} . \tag{2.16}$$

From (2.11) it follows that

$$N_{j+1} + N_{\sigma_{j+1}}(\xi) \geq P_{\sigma_{j+1}} . \tag{2.17}$$

The relations (2.15), (2.16), (2.17) imply (b).

3) Let $k \geq j+3$, $\sigma_j \neq \sigma_{j+1}$ and $\sigma_j = \sigma_{j+2}$. One can see that ξ is a reducible path. Then

$$h(\tilde{\Gamma}, P) \leq h(\Gamma, P) + \frac{1}{N(T_z, P)}$$
$$\times \left(N_j - N_{\sigma_j}(\xi) - 2N_{\sigma_{j+1}}(\xi) - \sum_{i=j+2,\sigma_i=\sigma_{j+1}}^{k-1} N_i \right). \tag{2.18}$$

The relation (i) implies

$$N_j + N_{j+1} \leq P_{\sigma_{j+1}} . \tag{2.19}$$

From (2.11) it follows that

$$N_{j+1} + \sum_{i=j+2,\sigma_i=\sigma_{j+1}}^{k-1} N_i + N_{\sigma_{j+1}}(\xi) \geq P_{\sigma_{j+1}} . \tag{2.20}$$

The relations (2.18), (2.19), (2.20) imply (c).

4) Let $k \geq j+3$, $\sigma_j \neq \sigma_{j+1}$ and $\sigma_j \neq \sigma_{j+2}$. One can see that ξ is a reducible path. Then

$$h(\tilde{\Gamma}, P) \leq h(\Gamma, P) + \frac{1}{N(T_z, P)}$$
$$\times \left(N_j - 2N_{\sigma_j}(\xi) - N_{\sigma_{j+1}}(\xi) - \sum_{i=j+2,\sigma_i=\sigma_j}^{k-1} N_i \right). \tag{2.21}$$

The relation (i) implies that

$$N_j \leq \frac{1}{2} P_{\sigma_j} . \tag{2.22}$$

From (2.11) it follows that

$$N_j + \sum_{i=j+2,\sigma_i=\sigma_j}^{k-1} N_i + N_{\sigma_j}(\xi) \geq P_{\sigma_j} . \tag{2.23}$$

The relations (2.21), (2.22), (2.23) imply (c).

5) Let $k \geq j+3$, $\sigma_j = \sigma_{j+1} = \ldots = \sigma_m$, and $\sigma_m \neq \sigma_{m+1}$ for some $m \in \{j+1, \ldots, k-2\}$. One can see that ξ is a reducible path. Then

$$h(\tilde{\Gamma}, P) \le h(\Gamma, P) + \frac{\sum_{i=j}^{m} N_i}{N(T_z, P)} - \frac{(m-j)}{N(T_z, P)}$$

$$\times \left(\sum_{i=m+1, \sigma_i = \sigma_{m+1}}^{k-1} N_i - (m - j + 1) N_{1-\sigma_j}(\xi) - N_{\sigma_j}(\xi) \right). \tag{2.24}$$

The relation (i) implies $N_i \le P_{1-\sigma_j}/2$ for $i = j, \dots, m$. Summing these inequalities, we obtain

$$\sum_{i=j}^{m} N_i \le \frac{m - j + 1}{2} P_{1-\sigma_j}. \tag{2.25}$$

From (2.11) it follows that

$$\sum_{i=m+1, \sigma_i = \sigma_{m+1}}^{k-1} N_i + N_{1-\sigma_j}(\xi) \ge P_{1-\sigma_j}. \tag{2.26}$$

The relations (2.24), (2.25), (2.26) imply (c).

One can see that the cases 1-5 cover all possible combinations of j, k,σ_j,..., σ_{k-1}. Statement (b) of the lemma holds for each case and statement (c) holds for the cases in which ξ is a reducible path, so the lemma is proved.   □

Lemma 2.3. *Let U be a 2-valued information system, Ψ a weight function for U, z a problem over U and P a probability distribution for z. Let Γ be a decision tree for the problem z, that solves z and is optimal for Ψ, z and P. Then Γ is a reduced decision tree.*

Proof. Let $U = (A, F)$. Suppose that Γ is not a reduced decision tree. Let v be an inessential node in Γ such that the path ϕ from the root to the node v does not contain any other inessential nodes. One can see that $T_z \pi(\phi) \ne \emptyset$. Denote by f the attribute assigned to v. Then there exists a number $\sigma \in \{0, 1\}$ such that $T_z \pi(\phi) = T_z \pi(\phi)(f, \sigma)$. For $\delta = 0, 1$, denote Γ_δ the subtree whose root the edge $e(\Gamma, v, \delta)$ enters. If v is not the root of Γ, then denote r the edge that enters v and transform Γ so that the edge r enters the root of the subtree Γ_σ. Delete from Γ the node v, the edges $e(\Gamma, v, 0)$, $e(\Gamma, v, 1)$ and the subtree $\Gamma_{1-\sigma}$. Denote $\tilde{\Gamma}$ the resulted tree. One can see that $\tilde{\Gamma}$ is a decision tree for z. Let us prove that $\tilde{\Gamma}$ solves the problem z.

For an arbitrary row $\bar{d} \in T_z$, denote $\xi^{\bar{d}}$ the complete path in Γ such that $\bar{d} \in T_z \pi(\xi^{\bar{d}})$. Since $T_z \pi(\phi) = T_z \pi(\phi)(f, \sigma)$, the terminal node of the path $\xi^{\bar{d}}$ is not contained in $\Gamma_{1-\sigma}$ and was not removed by the transformation. Denote $\tilde{\xi}^{\bar{d}}$ the complete path in $\tilde{\Gamma}$ that ends in the same terminal node as $\xi^{\bar{d}}$.

If $v \notin \xi^{\bar{d}}$, then the paths $\xi^{\bar{d}}$ and $\tilde{\xi}^{\bar{d}}$ coincide, so $T_z \pi(\xi^{\bar{d}}) = T_z \pi(\tilde{\xi}^{\bar{d}})$ and $\Psi(\xi^{\bar{d}}) = \Psi(\tilde{\xi}^{\bar{d}})$. If $v \in \xi^{\bar{d}}$, then the path $\tilde{\xi}^{\bar{d}}$ is resulted from $\xi^{\bar{d}}$ by removing the node v and the edge e_σ. Then $\Psi(\xi^{\bar{d}}) = \Psi(\tilde{\xi}^{\bar{d}}) - \Psi(f)$. Since $T_z \pi(\phi) = T_z \pi(\phi)(f, \sigma)$, the relation $T_z \pi(\xi^{\bar{d}}) = T_z \pi(\tilde{\xi}^{\bar{d}})$ holds. Then the fact that Γ solves z, implies that $\tilde{\Gamma}$ solves z. The relation $T_z \pi(\phi) \neq \emptyset$ implies $v \in \xi^{\bar{\delta}}$ for some $\bar{\delta} \in T_z$. Then $h_\Psi(\tilde{\Gamma}, P) \leq h_\Psi(\Gamma, P) - \Psi(f) P(\bar{\delta})/N(T_z, P)$ and $h_\Psi(\tilde{\Gamma}, P) < h_\Psi(\Gamma, P)$. The latter equality contradicts optimality of the tree Γ and the resulted contradiction proves the lemma. $\qquad\square$

Lemma 2.4. *Let z be a problem over an information system, P a probability distribution for z, and D a proper decomposition of the pair (z, P). Then there exists a decision tree for the problem z that solves z and is optimal for z and P, in which each path is ordered by basic attributes.*

Proof. Let Γ be a decision tree for z that solves z and is optimal for z and P. Denote $W(\Gamma)$ the set of nodes in Γ such that any node $w \in W(\Gamma)$ is assigned with a basic attribute and at least one node in the path $path(\Gamma, w)$ is assigned with an extended attribute. Obviously, if $W(\Gamma) = \emptyset$, then all paths in Γ are ordered by basic attributes.

Let $W(\Gamma) \neq \emptyset$. Consider an arbitrary complete path ξ in Γ that is not ordered by basic attributes. Let w be the first node in ξ that is contained in $W(\Gamma)$. According to Lemma 2.3, Γ is a reduced decision tree. Let us apply to ξ the path reduction operation and denote the resulted tree $\tilde{\Gamma}$. Lemma 2.2 implies that $\tilde{\Gamma}$ is a decision tree for z that solves z and is optimal for z and P. One can see that $W(\tilde{\Gamma}) \subseteq W(\Gamma) \setminus \{w\}$. Let us apply the above-mentioned transformation to $\tilde{\Gamma}$ and repeat this procedure until $W(\tilde{\Gamma}) = \emptyset$. Thus the desired decision tree is obtained in a finite number of steps. $\qquad\square$

Let z be a problem over an information system, P a probability distribution for z and $((z_0, P_0), (z_1, P_1), \ldots, (z_m, P_m))$ a proper decomposition for the pair (z, P). We will say that the tree Γ *is completely ordered* if for each row $\bar{d} \in T_{z_0}$, there exists a node $v^{\bar{d}}$ in Γ such that $T_{z_0}(path(\Gamma, v^{\bar{d}})) = \{\bar{d}\}$, and all nodes in the path $path(\Gamma, v^{\bar{d}})$ are assigned with basic attributes with possible exception of $v^{\bar{d}}$.

Lemma 2.5. *Let z be a problem over information system, P a probability distribution for z, and D a proper decomposition for the pair (z, P). Then there exists a decision tree for the problem z that solves z, is optimal for z and P, and is completely ordered.*

Proof. Let $D = ((z_0, P_0), (z_1, P_1), \ldots, (z_m, P_m))$, and $z_i = (\nu_i, f_1^i, \ldots, f_{n_i}^i)$ for $i = 0, \ldots, m$. Let $A_1, \ldots, A_m$ be the equivalence classes of the problem z_0, $T_{z_0} = \{\bar{\delta}^1, \ldots, \bar{\delta}^m\}$, and $\bar{\delta}^i = (f_1^0(a_i), \ldots, f_{n_0}^0(a_i))$, $a_i \in A_i$ for $i = 1, \ldots, m$.

According to Lemma 2.4, there exists a decision tree Γ for the problem z that solves z and is optimal for z and P, in which each complete path is ordered by basic attributes. We will say that rows $\bar{\delta}^i, \bar{\delta}^j \in T_{z_0}$, $i \neq j$ *are separated in Γ with basic attributes* if there is a node $v \in \Gamma$ that is assigned with a basic attribute f_k^0, and for some number $\delta \in \{0,1\}$, the relations $\bar{\delta}^i \in T_{z_0}\pi(path(\Gamma,v))(f_k^0,\delta)$ and $\bar{\delta}^j \in T_{z_0}\pi(path(\Gamma,v))(f_k^0,1-\delta)$ hold. Denote $R(\Gamma)$ the number of unordered pairs of rows in the table T_{z_0}, which are not separated in Γ with basic attributes. Obviously, if $R(\Gamma) = 0$, then Γ is a completely ordered decision tree.

Let $R(\Gamma) \neq 0$. Let $\bar{\delta}^i = (\delta_1^i, \ldots, \delta_{n_0}^i)$ and $\bar{\delta}^j = (\delta_1^j, \ldots, \delta_{n_0}^j)$ be rows of T_{z_0}, which are not separated in Γ with basic attributes. Choose a number $r \in \{1, \ldots, n_0\}$ such that $\delta_r^i \neq \delta_r^j$. Consider a row $\bar{d} = (d_1^0, \ldots, d_{n_0}^0, \ldots, d_1^m, \ldots, d_{n_m}^m) \in T_z$ for which $(d_1^0, \ldots, d_{n_0}^0) = \bar{\delta}^i$. Find a complete path ϕ in the tree Γ such that $\bar{d} \in T_z\pi(\phi)$. The inequality (ii) implies that there exists a row $\bar{c} = (c_1^0, \ldots, c_{n_0}^0, \ldots, c_1^m, \ldots, c_{n_m}^m) \in T_z$ such that $(c_1^0, \ldots, c_{n_0}^0) = \bar{\delta}^i$ or $(c_1^0, \ldots, c_{n_0}^0) = \bar{\delta}^j$, but $\bar{c} \notin T_z\pi(\phi)$. Then at least one node in the path ϕ is assigned with an extended attribute. Denote v_1 the first node of the path ϕ that is assigned with an extended attribute, and denote G the subtree of Γ whose root is v_1. Denote ξ a complete path in G such that each edge is assigned with the number 0. Let $\xi = v_1, e_1, \ldots, v_t, e_t, v_{t+1}$ where $v_1, \ldots, v_t \in V(\Gamma)$ and $e_1, \ldots, e_t \in E(\Gamma)$. Consider two cases.

1) Let $T_z\pi(path(\Gamma, v_{t+1})) = T_z\pi(path(\Gamma, v_{t+1}))(f_r^0, \sigma)$ for some $\sigma \in \{0,1\}$. Denote k the minimum number, for which the relation $T_z\pi(path(\Gamma, v_k)) = T_z\pi(path(\Gamma, v_k))(f_r^0, \sigma)$ holds. Denote $\bar{e}_{k-1} = e(\Gamma, v_{k-1}, 1)$, and denote w the node, which the edge $\bar{e}_{k-1}$ enters. Assign the attribute f_r^0 to the node v_{k-1}, the number σ to the edge e_{k-1} and the number $(1-\sigma)$ to the edge $\bar{e}_{k-1}$. Denote $\tilde{\Gamma}$ the resulted decision tree. In order to show that $\tilde{\Gamma}$ solves z, it is sufficient to prove correctness of the equality $T_z\pi(path(\Gamma, w)) = T_z\pi(path(\Gamma, w))(f_r^0, 1-\sigma)$. Since the node v_1 is assigned with an extended attribute and the path ξ is ordered by basic attributes, the node v_{k-1} is assigned with an extended attribute. Then $T_z\pi(path(\Gamma, w)) = T_z\pi(path(\Gamma, w))(f_r^0, \sigma_1)$ for some $\sigma_1 \in \{0,1\}$. By the choice of the number k, the relation

$$T_z\pi(path(\Gamma, v_{k-1})) \neq T_z\pi(path(\Gamma, v_{k-1}))(f_r^0, \sigma)$$

holds. Therefore,

$$T_z\pi(path(\Gamma, w)) = T_z\pi(path(\Gamma, w))(f_r^0, 1-\sigma),$$

and $\tilde{\Gamma}$ solves the problem z. Obviously, $h(\tilde{\Gamma}, P) = h(\Gamma, P)$ and the tree $\tilde{\Gamma}$ is optimal for z and P. Denote $\tilde{\xi}$ the complete path in the tree $\tilde{\Gamma}$ that ends in the node v_{t+1}. Let us apply to $\tilde{\xi}$ the path reduction operation and denote the resulted tree $\hat{\Gamma}$. From Lemma 2.2 it follows that $\hat{\Gamma}$ is a decision tree for the problem z that solves z and is optimal for z and P. One can see that each complete path in $\hat{\Gamma}$ is ordered by basic attributes, and $R(\hat{\Gamma}) \leq R(\Gamma) - 1$.

2) Let $T_z \pi(path(\Gamma, v_{t+1})) \neq T_z \pi(path(\Gamma, v_{t+1}))(f_r^0, \sigma)$ for $\sigma = 0, 1$. Then the inequalities (i) and (ii) from the definition of the proper problem decomposition imply that $t \geq 3$, and for some $k_1, k_2 \in \{1, \ldots, t\}$, the node v_{k_1} is assigned with an attribute from the set $\{\tilde{f}_1^i, \ldots, \tilde{f}_{n_i}^i\}$, and the node v_{k_2} is assigned with an attribute from the set $\{\tilde{f}_1^j, \ldots, \tilde{f}_{n_j}^j\}$. Denote α the number assigned to the node v_{t+1}. Assign the attribute f_r^0 to the node v_{t+1}, add two edges leaving this node and label them with the numbers 0 and 1 respectively. Add to the tree Γ two nodes w_0 and w_1, assign the number α to these nodes and transform the tree Γ so that the edge $e(\Gamma, v_{t+1}, \sigma)$ enters the node w_σ for $\sigma = 0, 1$. Denote the resulted tree $\tilde{\Gamma}$. One can see that $\tilde{\Gamma}$ is a decision tree for the problem z that solves z and $h(\tilde{\Gamma}, P) = h(\Gamma, P) + N(T_z\pi(\xi), P)/N(T_z, P)$. Denote $\tilde{\xi}$ the complete path in the tree $\tilde{\Gamma}$ that ends in w_0. One can see that $N_0(\tilde{\xi}) + N_1(\tilde{\xi}) = N(T_z\pi(\xi), P)$. Apply the path reduction operation to the path $\tilde{\xi}$ and denote the resulted tree $\hat{\Gamma}$. From Lemma 2.2 it follows that $\hat{\Gamma}$ is a decision tree for the problem z that solves z, and $h(\hat{\Gamma}, P) \leq h(\tilde{\Gamma}, P) - (N_0(\tilde{\xi}) + N_1(\tilde{\xi}))/N(T_z, P)$. Then $h(\hat{\Gamma}, P) \leq h(\Gamma, P)$ and the decision tree $\hat{\Gamma}$ is optimal for z and P. One can see that each complete path in the tree $\hat{\Gamma}$ is ordered by basic attributes and $R(\hat{\Gamma}) \leq R(\Gamma) - 1$.

Let us apply the above-mentioned transformation to $\hat{\Gamma}$ and repeat this procedure until $R(\Gamma) = 0$. Thus we obtain the desired decision tree in a finite number of steps.   □

Proof of Theorem 2.5. Let us show that the decision tree $\Phi = \Phi(\Gamma_0, \Gamma_1, \ldots, \Gamma_m)$ solves the problem z. Denote by $A_1, \ldots, A_m$ the equivalence classes of the problem z_0. Let $T_{z_0} = \{\bar{d}^1, \ldots, \bar{d}^m\}$ where $\bar{d}^i = (f_1^0(a_i), \ldots, f_{n_0}^0(a_i))$, $a_i \in A_i$, $i = 1, \ldots, m$. Consider an arbitrary row $\bar{\delta} = (\delta_1^0, \ldots, \delta_{n_0}^0, \ldots, \delta_1^m, \ldots, \delta_{n_m}^m) \in T_z$. Let $(\delta_1^0, \ldots, \delta_{n_0}^0) = \bar{d}^i$ for some $i \in \{1, \ldots, m\}$. Denote $\xi^{\bar{\delta}}$ the complete path in the decision tree Φ such that $\bar{\delta} \in T_z\pi(\xi^{\bar{\delta}})$. From the definition of the tree Φ it follows that the terminal node of the path $\xi^{\bar{\delta}}$ belongs to the subtree $\hat{\Gamma}_i$. Since the decision tree Γ_i solves the problem z_i, the terminal node of the path ξ_i is assigned with the number $\nu_i(\delta_1^i, \ldots, \delta_{n_i}^i)$. From the definition of proper decomposition we have $\nu(\bar{\delta}) = \nu_i(\delta_1^i, \ldots, \delta_{n_i}^i)$. Therefore, Φ solves the problem z.

Let us show that Φ is optimal for z and P. From Lemma 2.5 it follows that there exists a decision tree G for the problem z that solves z, is optimal for z and P, and is completely ordered. For $i = 1, \ldots, m$, denote v_i the node of the tree G such that $T_{z_0} \pi(path(G, v_i)) = \{\bar{d}^i\}$ and all nodes in the path $path(G, v_i)$, are assigned with basic attributes with possible exception of the node v_i. For $i = 1, \ldots, m$, delete from the tree G the subtree whose root is v_i, but leave v_i itself. Assign to v_i the number $\nu_0(\bar{d}^i)$. Denote the resulted tree G_0. According to Lemma 2.3, G is a reduced decision tree. It implies that all nonterminal nodes of the tree G_0 are encountered in the paths $path(G_0, v_1), \ldots, path(G_0, v_m)$. Then all nonterminal nodes in G_0 are assigned with basic attributes and G_0 is a decision tree for z_0. One can see that G_0 solves z_0. Then,

$$h(G_0, P_0) \geq h(z_0, P_0) . \tag{2.27}$$

Let $z_i = (\nu_i, f_1^i, \ldots, f_{n_i}^i)$ for $i = 0, \ldots, m$, and $z = (\nu, f_1^0, \ldots, f_{n_0}^0, \tilde{f}_1^1, \ldots, \tilde{f}_{n_1}^1, \ldots, \tilde{f}_1^m, \ldots, \tilde{f}_{n_m}^m)$.

For an arbitrary $i \in \{1, \ldots, m\}$, consider the subtree $\tilde{G}_i$ of the tree G whose root is the node v_i. By definition, $\tilde{f}_k^j \equiv 0$ in the set A_i for any $j \in \{1, \ldots, m\} \setminus \{i\}$, $k \in \{1, \ldots, n_j\}$. Then the fact that G is a reduced decision tree implies that all nonterminal nodes of the tree $\tilde{G}_i$ are assigned with attributes from the set $\{\tilde{f}_1^i, \ldots, \tilde{f}_{n_i}^i\}$. For each nonterminal node w in $\tilde{G}_i$, let us replace the attribute $\tilde{f}_j^i$ assigned to w with the corresponding attribute f_j^i. Denote G_i the resulted tree. One can see that G_i is a decision tree for z_i that solves z_i. Then

$$h(G_i, P_i) \geq h(z_i, P_i) . \tag{2.28}$$

Let us compare the average depth of the trees G and Φ. One can see that

$$h(\Phi, P) = h(\Gamma_0, P_0) + \frac{1}{N(T_z, P)} \sum_{i=1}^{m} N(T_{z_i}, P_i) h(\Gamma_i, P_i)$$

and

$$h(G, P) = h(G_0, P_0) + \frac{1}{N(T_z, P)} \sum_{i=1}^{m} N(T_{z_i}, P_i) h(G_i, P_i) .$$

Since $\Gamma_0, \Gamma_1, \ldots, \Gamma_m$ are optimal decision trees, the inequalities (2.27) and (2.28) result in $h(\Phi, P) \leq h(G, P)$. Therefore, Φ is an optimal decision tree for z and P. $\qquad \square$

2.4.3 Example of Decomposable Problem

Theorem 2.5 allows for finding decision trees with the minimum average depth for some classes of problems. This section shows that the upper bound on the average depth of decision tree given by Theorem 2.4 is close to unimprovable.

Theorem 2.6. *For arbitrary natural numbers $m \geq 2$, n, there exists a 2-valued information system U_m^n, a problem z_m^n over U_m^n with m^n equivalence classes and a probability distribution $P_m^n \equiv 1$ such that $H(P) = n\log_2 m$,*

$$
M(z) = \begin{cases} m - 1, & \text{if } n = 1, \\ m, & \text{if } n = 2, \\ m + 1, & \text{if } n \geq 3, \end{cases} \quad \text{and } h(z,P) = \frac{(m+2)(m-1)}{2m} n .
$$

We preface the proof of the theorem by several auxiliary definitions. Let $m \geq 2$, n be arbitrary natural numbers. Define a system of circles B_m^n in a plane. By definition, B_m^1 is m non-intersecting circles such that no one is enclosed to another. Let the system B_m^{i-1} have been already defined. Then the system B_m^i consists of m non-intersecting circles, such that no one is enclosed to another, and there is a system of the kind B_m^{i-1} inside each circle. A circle in B_m^n is called *zero order* circle if it does not contain any circles from B_m^n. Let for some $i < n$, circles of orders from zero to $(i-1)$ have been already defined. A circle from B_m^n is called *i-th order circle* if the order of all enclosed circles is at most $(i - 1)$ and is equal to $(i - 1)$ for at least one circle. One can see that B_m^n contains $s = m^n$ zero order circles. Denote these circles $C_1, \ldots, C_s$. For $i = 1, \ldots, s$, denote a_i a point inside C_i, and denote $A = \{a_1, \ldots, a_s\}$. Set into correspondence to each circle C from B_m^n a function $f : A \to \{0,1\}$. The function f takes the value 1 on an element a_i if the point a_i is located inside the circle C, and 0 otherwise. Denote $F = \{f_1, \ldots, f_t\}$ the set of functions that correspond to all circles from B_m^n. Then $U_m^n = (A, F)$.

Let $z_m^n = (\nu, f_1, \ldots, f_t)$ be a diagnostic problem over U_m^n. The following lemma gives the value of the parameter $M(z)$ for the problem z_m^n.

Lemma 2.6. *Let $m \geq 2, n$ be arbitrary natural numbers. Then*

$$
M(z_m^n) = \begin{cases} m - 1, & \text{if } n = 1, \\ m, & \text{if } n = 2, \\ m + 1, & \text{if } n \geq 3. \end{cases}
$$

Proof. Consider the case $n \geq 3$. Let us calculate $M(z, \bar{\delta})$ for an arbitrary tuple $\bar{\delta} \in \{0,1\}^t$. Let $\bar{\delta} = (0, \ldots, 0)$. The system of equations $\{f_1(x) = 0, \ldots, f_m(x) = 0\}$ does not have a solution on the set A if $f_1, \ldots, f_m$ are pairwise different attributes corresponding to $(n-1)$-th order circles. Therefore, $M(z_m^n, \bar{\delta}) \leq m$. Let $\bar{\delta} \neq (0, \ldots, 0)$. Denote C_0 a circle of the smallest order such that its corresponding attribute takes the value 1 on $\bar{\delta}$. Denote n_0 the order of the circle C_0 and f_{i_0} its corresponding attribute. If $n_0 = 0$, then the equation $f_{i_0}(x) = 1$ has a single solution on the set A, and $M(z_m^n, \bar{\delta}) = 1$. Let $n_0 \geq 1$. Denote $f_{i_1}, \ldots, f_{i_m}$ the attributes corresponding to the $(n_0 - 1)$-th order circles that are enclosed into C_0. By the choice of C_0, the attributes $f_{i_1}, \ldots, f_{i_m}$ take the value 0 on $\bar{\delta}$. The system of equations $\{f_{i_0}(x) = 1, f_{i_1}(x) = 0, \ldots, f_{i_m}(x) = 0\}$ does not have a solution on the set A, so $M(z_m^n, \bar{\delta}) \leq m + 1$. Therefore,

$$M(z_m^n) \leq m + 1 . \tag{2.29}$$

Let C_2 be an arbitrary second order circle. Consider a tuple $\bar{\delta} = (\delta_1, \ldots, \delta_t)$ in which the values of the attributes corresponding to C_2 and all circles that contains C_2 are set to 1, and all other elements are set to 0. Let us show that $M(z_m^n, \bar{\delta}) \geq m + 1$. Denote f_{i_0} the attribute corresponding to C_2, and $f_{i_1}, \ldots, f_{i_m}$ the attributes corresponding to the first order circles enclosed into C_2. Let $S = \{f_{j_1}(x) = \delta_{j_1}, \ldots, f_{j_k}(x) = \delta_{j_k}\}$ be an arbitrary system of equations that either does not have a solution on the A or $z_m^n(x) \equiv \text{const}$ on the set of solutions. Let $l \in \{1, \ldots, k\}$. Replace the equation $f_{j_l}(x) = \delta_{j_l}$ with the equation $f_{i_r}(x) = 0$ if the circle corresponding to the attribute f_{j_l} is either enclosed into the circle corresponding to f_{i_r} or coincides with it for some $r \in \{1, \ldots, m\}$. Otherwise, replace the equation $f_{j_l}(x) = \delta_{j_l}$ with the equation $f_{i_0}(x) = 1$. Let us make such replacement for $l = 1, \ldots, s$, and denote the resulted system S_1. One can see that the system S_1 contains at most k equations, the set of solutions of S_1 on A is a subset of the set of solutions of S on A, and S_1 is a subsystem of the system $S_2 = \{f_{i_0}(x) = 1, f_{i_1}(x) = 0, \ldots, f_{i_m}(x) = 0\}$. Assume that $S_1 \neq S_2$. One can see that in this case $z_m^n(x) \not\equiv \text{const}$ on the set of solutions of S_1 on A. But this is impossible. Therefore, $k \geq m+1$. Then any system of equations that either do not have a solution on A or have a set of solutions coinciding with some equivalence class contains at least $(m + 1)$ equations, and $M(z_m^n, \bar{\delta}) \geq m + 1$. This inequality and (2.29) imply $M(z_m^n) = m + 1$. The cases $n = 1$ and $n = 2$ are considered similarly. $\square$

Proof of Theorem 2.6. We apply induction on n. Let $n = 1$. Define a decision tree Γ for the problem z_m^1. The decision tree Γ contains $(m - 1)$ nonterminal nodes $v_1, \ldots, v_{m-1}$, which are assigned with pairwise different attributes

$f_1, \ldots, f_{m-1}$, and m terminal nodes $v_m, w_1, \ldots, w_{m-1}$. For $i = 1, \ldots, m-1$, two edges leave the node v_i that are labeled with the numbers 0 and 1 respectively. The edge labeled with 0 enters the node v_{i+1} and the edge labeled with 1 enters the node w_i. For $i = 1, \ldots, m-1$, the node w_i is assigned with the number $z_m^1(a_i)$ where a_i is an element of the set A such that $f_i(a_i) = 1$. The node v_m is assigned with the number $z_m^1(a_0)$ where a_0 is an element of the set A such that $f_i(a_0) = 0$ for $i = 1, \ldots, m-1$. The decision tree Γ does not contain any other nodes and edges. One can see that Γ solves the problem z_m^1 and is optimal for z_m^1 and P_m^1. Let us calculate the average depth of Γ:

$$h(\Gamma, P_m^1) = \frac{1}{m}\left(\sum_{i=1}^{m-1} i + (m-1)\right) = \frac{(m+2)(m-1)}{2m} .$$

Then, $h(z_m^1, P_m^1) = (m+2)(m-1)/(2m)$.

Let n be a natural number greater than 1 such that the theorem holds for all natural numbers less than n. Denote $C_1, \ldots, C_m$ the $(n-1)$-th order circles contained in the system B_m^n, and $f_1, \ldots, f_m$ their corresponding attributes. Consider a decomposition $((z_0, P_0), (z_1, P_1), \ldots, (z_m, P_m))$ of the pair (z_m^n, P_m^n). The diagnostic problem z_0 contains only the attributes $f_1, \ldots, f_m$. For $i = 1, \ldots, m$, the problem z_i contains all attributes corresponding to the circles enclosed in C_i, and $z_i(x)$ is the mapping $z_m^n : A \to \omega$ restricted to the set $\{a \in A : f_i(a) = 1\}$. Let for $i = 0, \ldots, m$, P_i be a uniform probability distribution for the problem z_i. One can see that $((z_0, P_0), (z_1, P_1), \ldots, (z_m, P_m))$ is a proper decomposition of the pair (z_m^n, P_m^n), $h(z_0, P_0) = h(z_m^1, P_m^1)$ and $h(z_i, P_i) = h(z_m^{n-1}, P_m^{n-1})$ for $i = 1, \ldots, m$. Using induction hypothesis, we obtain $h(z_0, P_0) = (m+2)(m-1)/(2m)$ and $h(z_i, P_i) = [(m+2)(m-1)/(2m)](n-1)$ for $i = 1, \ldots, m$. Let Γ_i be a decision tree for the problem z_i that solves z_i and is optimal for z_i and P_i, $i = 0, \ldots, m$. Let $\Phi = \Phi(\Gamma_0, \Gamma_1, \ldots, \Gamma_m)$. From the definition of the tree Φ it follows that

$$h(\Phi, P_m^n) = h(\Gamma_0, P_0) + \sum_{i=1}^{m} \frac{h(\Gamma_i, P_i)}{m} = \frac{(m+2)(m-1)}{2m}n .$$

Using Theorem 2.5, we obtain $h(z_m^n, P_m^n) = [(m+2)(m-1)/(2m)]n$. $\square$

Chapter 3
Representing Boolean Functions by Decision Trees

A Boolean or discrete function can be represented by a decision tree. A compact form of decision tree named binary decision diagram or branching program is widely known in logic design [2, 40]. This representation is equivalent to other forms, and in some cases it is more compact than values table or even the formula [44]. Representing a function in the form of decision tree allows applying graph algorithms for various transformations [10]. Decision trees and branching programs are used for effective hardware [15] and software [5] implementation of functions. For the implementation to be effective, the function representation should have minimal time and space complexity. The average depth of decision tree characterizes the expected computing time, and the number of nodes in branching program characterizes the number of functional elements required for implementation. Often these two criteria are incompatible, i.e. there is no solution that is optimal on both time and space complexity.

The chapter considers several problems of representing functions in the form of decision trees. It consists of two sections. The first section studies the average time complexity of representing Boolean functions by decision trees. The complexity of a class of functions can be characterized by a Shannon type function $H(n)$ that shows the dependence of the minimum average depth of decision tree in the worst case on the number of function arguments. For each closed class of Boolean functions B, a lower and an upper bound on $H_B(n)$ are given. Analogous results for the depth of decision trees are described in [48]. The second section considers branching programs with the minimum average weighted depth. It is proven that such programs are read-once, i.e. each attribute is checked at most once along each path. This fact implies high lower bounds on the number of nodes in branching programs with the minimum average weighted depth for several known functions.

Results of this chapter were previously published in [19].

I. Chikalov: Average Time Complexity of Decision Trees, ISRL 21, pp. 41–60.
springerlink.com

3.1   On Average Depth of Decision Trees Implementing Boolean Functions

This section contains some auxiliary notions followed by propositions that give bounds on the function $H(n)$ for all closed classes of Boolean functions. The notation of closed classes of Boolean functions is in accordance with [36]; the classes and the class inclusion diagram are described in Appendix A.

3.1.1   Auxiliary Notions

A function of the form $f : E_2^n \to E_2$ is called *Boolean function*. The constants 0 and 1 also are Boolean functions.

Let $f(x_1, \ldots, x_n)$ be a Boolean function. A variable x_i of the function f will be called *essential* if there exist two n-tuples $\bar{\delta}$ and $\bar{\sigma}$ from E_2^n which differ only in the i-th digit and for which $f(\bar{\delta}) \neq f(\bar{\sigma})$. Variables of the function f which are not essential will be called *inessential*.

Let us set into correspondence to a Boolean function $f(x_1, \ldots, x_n)$ a problem $z = (f, x_1, \ldots, x_n)$ over the information system $U_n = (E_2^n, \{x_1, \ldots, x_n\})$. The problem z has two equivalence classes Q_0 and Q_1 containing the sets of binary tuples on which f takes the value 0 and 1 respectively. A decision tree solving the problem z is called *a decision tree implementing f*. Denote $g(f)$ and $h(f)$ respectively the minimum depth of a decision tree implementing f and the minimum average depth of a decision tree implementing f relative to the probability distribution $P \equiv 1$.

Denote $\dim f$ the number of arguments of the function f. Let B be a set of Boolean functions. Consider the functions

$$\mathcal{G}_B(n) = \max\{g(f) : f \in B, \dim f \leq n\}$$

and

$$\mathcal{H}_B(n) = \max\{h(f) : f \in B, \dim f \leq n\} \quad .$$

that characterize the growth in the worst case of the minimum depth and the minimum average depth of decision trees implementing Boolean functions from B with growth of the number of function arguments. Note that $\mathcal{H}_B(n) \leq \mathcal{G}_B(n)$ for any n.

3.1.2   Bounds on Function $\mathcal{H}_B(n)$

In this section, a number of propositions are formulated that for each closed class of Boolean functions B give the upper and the lower bound on $\mathcal{H}_B(n)$ followed by two theorems that compare the values $\mathcal{G}_B(n)$ and $\mathcal{H}_B(n)$.

Proposition 3.1. *For $B \in \{O_2, O_3, O_7\}$, the relation $\mathcal{H}_B(n) = 0$ holds.*

Proposition 3.2. *For $B \in \{O_1, O_4, O_5, O_6, O_8, O_9\}$, the relation $\mathcal{H}_B(n) = 1$ holds.*

Proposition 3.3. *For $B \in \{S_1, S_3, S_5, S_6, P_1, P_3, P_5, P_6\}$, the relation*

$$\mathcal{H}_B(n) = \begin{cases} 2 - \frac{1}{2^{n-1}}, & \text{if } n \geq 2, \\ 1, & \text{if } n = 1. \end{cases}$$

holds.

Proposition 3.4. *For $B \in \{L_1, L_2, L_3, C_1, C_2, C_3\}$, the relation $\mathcal{H}_B(n) = n$ holds.*

Proposition 3.5. *For $B \in \{L_4, L_5\}$, the relation*

$$\mathcal{H}_B(n) = \begin{cases} n, & \text{if } n = 2k + 1, \, k \geq 0, \\ n - 1, & \text{if } n = 2k, \, k \geq 1 \end{cases}$$

holds.

Proposition 3.6. *For $B = C_4$, the relation*

$$\mathcal{H}_B(n) = \begin{cases} n, & \text{if } n = 2k + 1, \, k \geq 0, \\ n - \frac{1}{2^{n-1}}, & \text{if } n = 2k, \, k \geq 1 \end{cases}$$

holds.

Proposition 3.7. *For $B \in \{D_1, D_3\}$, the following relations hold:*
a) $\mathcal{H}_B(n) = n$, if $n = 2k + 1$, $k \geq 0$;
b) $n - 1.7/\sqrt{n} \leq \mathcal{H}_B(n) \leq n - 1/2^{n-1}$, if $n = 2k$, $k \geq 1$.

Proposition 3.8. *For $B \in \{M_1, M_2, M_3, M_4\}$, the relation*

$$n + 1 - \sqrt{n+1} \leq \mathcal{H}_B(n) \leq n - \lfloor n/2 \rfloor \, 2^{-\lfloor n/2 \rfloor}$$

holds.

Proposition 3.9. *For $B = D_2$, the relation*

$$n + 1/2 - \sqrt{n+1} \le \mathcal{H}_B(n) \le n - \lfloor n/2 \rfloor 2^{-\lfloor n/2 \rfloor}$$

holds.

Proposition 3.10. *For $B \in \{F_1^\infty, F_4^\infty, F_5^\infty, F_8^\infty\}$, the relation $\mathcal{H}_B(n) = (n+1)/2$ holds.*

Proposition 3.11. *For $B \in \{F_2^\infty, F_3^\infty, F_6^\infty, F_7^\infty\}$, the relation*

$$1 + (n - \sqrt{n})/2 \le \mathcal{H}_B(n) \le (n+1)/2, \ n \ge 1$$

holds.

Proposition 3.12. *For $B \in \{F_1^\mu, F_4^\mu, F_5^\mu, F_8^\mu\}$, $\mu \ge 2$, the relation*

$$(n+1)/2 \le \mathcal{H}_B(n) \le n - \lfloor n/2 \rfloor 2^{-\lfloor n/2 \rfloor}$$

holds.

Proposition 3.13. *For any $B \in \{F_2^\mu, F_3^\mu, F_6^\mu, F_7^\mu\}$, $\mu \ge 2$, the relation*

$$1 + (n - \sqrt{n})/2 \le \mathcal{H}_B(n) \le n - \lfloor n/2 \rfloor 2^{-\lfloor n/2 \rfloor}$$

holds.

The following theorem is proved by Moshkov.

Theorem 3.1 ([48]). *Let B be a closed class of Boolean functions and n be a natural number. Then*

a) if $B \in \{O_2, O_3, O_7\}$, then $\mathcal{G}_B(n) = 0$;
b) if $B \in \{O_1, O_4, O_5, O_6, O_8, O_9\}$, then $\mathcal{G}_B(n) = 1$;

c) if $B \in \{L_4, L_5\}$, then $\mathcal{G}_B(n) = \begin{cases} n, & \text{if n is odd}, \\ n-1, & \text{if n is even}; \end{cases}$

d) if $B \in \{D_1, D_2, D_3\}$, then $\mathcal{G}_B(n) = \begin{cases} n, & \text{if $n \ge 3$}, \\ 1, & \text{if $n \le 2$}; \end{cases}$

e) if the class B does not coincide with any classes mentioned in (a) – (d), then $\mathcal{G}_B(n) = n$.

The following two theorems immediately follow from the previous theorem and Propositions 3.1-3.13. These theorems characterize the relation between $\mathcal{H}_B(n)$ and $\mathcal{G}_B(n)$ for each closed class of Boolean functions.

Theorem 3.2. *Let B be a closed class of Boolean functions, and n a natural number. Let at least one of the following conditions hold:*

a) $n = 1$;
b) $B \in \{O_1, \ldots, O_9, L_1, \ldots, L_5, C_1, C_2, C_3\}$;
c) $B \in \{C_4, D_1, D_3\}$ and n is odd;
d) $B \in \{D_1, D_2, D_3\}$ and $n = 2$.

Then $\mathcal{H}_B(n) = \mathcal{G}_B(n)$. If none of the conditions (a), (b), (c), (d) hold, then $\mathcal{H}_B(n) < \mathcal{G}_B(n)$.

Theorem 3.3. *Let B be a closed class of Boolean functions. Then*

a) $\lim_{n\to\infty} \mathcal{H}_B(n)/\mathcal{G}_B(n) = 0$ if $B \in \{S_1, S_3, S_5, S_6, P_1, P_3, P_5, P_6\}$;
b) $\mathcal{H}_B(n)/\mathcal{G}_B(n) = 1$ if $B \in \{O_1, \ldots, O_9, L_1, \ldots, L_5, C_1, C_2, C_3\}$;
c) $\lim_{n\to\infty} \mathcal{H}_B(n)/\mathcal{G}_B(n) = 1$ if $B \in \{C_4, M_1, \ldots, M_4, D_1, D_2, D_3\}$;
d) $\lim_{n\to\infty} \mathcal{H}_B(n)/\mathcal{G}_B(n) = 1/2$ if $B \in \{F_1^\infty, \ldots, F_8^\infty\}$;
e)

$$\frac{1}{2} - \varepsilon(n) < \frac{\mathcal{H}_B(n)}{\mathcal{G}_B(n)} < 1$$

where $\varepsilon(n) = O(1/\sqrt{n})$ if $B \in \{F_1^\mu, \ldots, F_8^\mu\}$ and $\mu \geq 2$.

3.1.3   Proofs of Propositions 3.1-3.13

We preface proof of the propositions by a series of lemmas. Since in this section the uniform probability distribution is assumed, it is omitted in notations, so the average depth of a tree Γ is denoted by $h(\Gamma)$.

For an arbitrary path ξ in a decision tree Γ, denote its length by $l_\Gamma(\xi)$. Denote the logical negation operation by $\neg$ and the modulo 2 summation by $\oplus$. A Boolean function $f(x_1, \ldots, x_n)$ is called *symmetrical* if for each tuple $\bar{\delta} \in E_2^n$ and each permutation p of n elements, the relation $f(\bar{\delta}) = f(p(\bar{\delta}))$ holds.

Lemma 3.1. *Let $f_0(x_1, \ldots, x_n)$ and $f_1(x_1, \ldots, x_n)$ be arbitrary Boolean functions for some natural number n. Let Γ_0 and Γ_1 be decision trees implementing f_0 and f_1 respectively. Let Γ be a decision tree of the following form:*

a) the root of Γ is assigned with the attribute x_{n+1};
b) for $\delta = 0, 1$, an edge e_δ leaves the root of Γ and enters the root of Γ_δ, which is labeled with the number δ;
c) Γ does not contain any other nodes and edges.

Then the following statements are true:

a) *the decision tree Γ implements the function* $\neg x_{n+1} \wedge f_0 \vee x_{n+1} \wedge f_1$;

b) $h(\Gamma) = 1 + (h(\Gamma_0) + h(\Gamma_1))/2$.

Proof. Consider an arbitrary tuple $\bar{\delta} = (\delta_1, \ldots, \delta_{n+1})$ and find in the tree Γ the path $\xi(\bar{\delta})$ on which computations for $\bar{\delta}$ are performed. Since the root of Γ is assigned with the attribute x_{n+1}, the terminal node of the path $\xi(\bar{\delta})$ is located in the tree $\Gamma_{\delta_{n+1}}$. Denote $\bar{\delta}^* = (\delta_1, \ldots, \delta_n)$ and denote $\xi^{\delta_{n+1}}(\bar{\delta}^*)$ the part of $\xi(\bar{\delta})$ from the root of the tree $\Gamma_{\delta_{n+1}}$ to the terminal node. One can see that in the tree $\Gamma_{\delta_{n+1}}$, computations for the tuple $\bar{\delta}^*$ are performed along the path $\xi^{\delta_{n+1}}(\bar{\delta}^*)$. Since the tree $\Gamma_{\delta_{n+1}}$ implements the function $f_{\delta_{n+1}}$, the terminal node of the path $\xi(\bar{\delta})$ is assigned with the number $f_{\delta_{n+1}}(\delta_1, \ldots, \delta_n) = \neg \delta_{n+1} \wedge f_0(\delta_1, \ldots, \delta_n) \vee \delta_{n+1} \wedge f_1(\delta_1, \ldots, \delta_n)$. Therefore, the tree Γ implements the function $\neg x_{n+1} \wedge f_0 \vee x_{n+1} \wedge f_1$.

Obviously, the length of the path $\xi(\bar{\delta})$ is greater by 1 than the length of the path $\xi^{\delta_{n+1}}(\bar{\delta}^*)$. Then

$$h(\Gamma) = \frac{1}{2^{n+1}} \sum_{\delta \in E_2^{n+1}} l_\Gamma(\xi(\delta))$$

$$= \frac{1}{2} \left(\frac{1}{2^n} \sum_{\bar{\delta}^* \in E_2^n} (l_\Gamma(\xi^0(\bar{\delta}^*)) + 1) + \frac{1}{2^n} \sum_{\bar{\delta}^* \in E_2^n} (l_\Gamma(\xi^1(\bar{\delta}^*)) + 1) \right)$$

$$= \frac{1}{2}(h(\Gamma_0) + 1 + h(\Gamma_1) + 1) = 1 + \frac{h(\Gamma_0) + h(\Gamma_1)}{2}.$$

$\square$

The following lemma gives a combinatorial identity, which will be used further. Denote $g(n,t) = \sum_{i=t}^n C_i^t / 2^i$.

Lemma 3.2. *For arbitrary natural numbers $n, t \leq n$, the equality $g(n,t) = 2 - 1/2^n \sum_{i=0}^t C_{n+1}^i$ holds.*

Proof. Apply the following transformations:

$$g(n,t) = \sum_{i=t}^n \frac{1}{2^i} C_i^t = \sum_{i=t}^{n+1} \frac{1}{2^i} C_i^t - \frac{1}{2^{n+1}} C_{n+1}^t = \sum_{i=t-1}^n \frac{1}{2^{i+1}} C_{i+1}^t$$

$$- \frac{1}{2^{n+1}} C_{n+1}^t = \frac{1}{2} \left(\sum_{i=t-1}^n \frac{1}{2^i} C_i^t + \sum_{i=t-1}^n \frac{1}{2^i} C_i^{t-1} - \frac{1}{2^n} C_{n+1}^t \right)$$

$$= \frac{1}{2} \left(g(n,t) + g(n,t-1) - \frac{1}{2^n} C_{n+1}^t \right).$$

Then we have

$$g(n,t) = g(n,t-1) - \frac{1}{2^n} C_{n+1}^t . \tag{3.1}$$

Let us modify $g(n,1)$ as follows:

$$g(n,1) = \sum_{i=1}^{n} \frac{i}{2^i} = \frac{1}{2} + \frac{2}{4} + \ldots + \frac{n-1}{2^{n-1}} + \frac{n}{2^n} = \frac{1}{2} + \frac{2}{4} + \ldots$$

$$+ \frac{n-1}{2^{n-1}} + \frac{n}{2^{n-1}} - \frac{n}{2^n} = 1 + \frac{1}{2} + \frac{1}{4} + \ldots + \frac{1}{2^{n-1}} - \frac{n}{2^n}$$

$$= \frac{1 - \frac{1}{2^n}}{1 - \frac{1}{2}} - \frac{n}{2^n} = 2\frac{2^n - 1}{2^n} - \frac{n}{2^n} = 2 - \frac{1}{2^n} - \frac{n+1}{2^n} .$$

Then $g(n,1)$ can be expressed in the following form:

$$g(n,1) = 2 - \frac{1}{2^n} \left(C_{n+1}^0 + C_{n+1}^1 \right) . \tag{3.2}$$

The equalities (3.1) and (3.2) imply $g(n,t) = 2 - 1/2^n \sum_{i=0}^{t} C_{n+1}^i$.   □

Lemma 3.3. *The function*

$$f(n) = (n+1) - \frac{\sqrt{2}(n+1)}{\sqrt{3n+5}} - (n + \frac{3}{2} - \sqrt{n+2})$$

takes positive values for any natural number $n \geq 3$.

Proof. Convert all terms to the common denominator:

$$f(n) = \sqrt{n+2} - \frac{1}{2} - \frac{\sqrt{2}(n+1)}{\sqrt{3n+5}}$$

$$= \frac{2\sqrt{n+2}\sqrt{3n+5} - \sqrt{3n+5} - 2\sqrt{2}(n+1)}{2\sqrt{3n+5}} = \frac{\phi(n)}{2\sqrt{3n+5}} .$$

The denominator is always positive, so $f(n)$ has the same sign as $\phi(n)$. Apply the following transformations:

$$\phi(n) = (4n+7) - (\sqrt{3n+5} - \sqrt{n+2})^2 - \sqrt{3n+5} - 2\sqrt{2}(n+1)$$

$$> (4n+7) - (\sqrt{3n+6} - \sqrt{n+2})^2 - \sqrt{3n+5} - 2\sqrt{2}(n+1)$$

$$= (4n+7) - (\sqrt{3} - 1)^2(n+2) - \sqrt{3n+5} - 2\sqrt{2}(n+1)$$

$$= 2(\sqrt{3} - \sqrt{2})n + 4\sqrt{3} - 2\sqrt{2} - 1 - \sqrt{3n+5} > 0.6n + 3 - \sqrt{3n+5}$$

$$= \psi(n) .$$

The facts that $\psi(n)$ is monotonically increasing and $\psi(3) > 0$ prove the lemma.   □

For an arbitrary natural number k, the Boolean function that takes value 1 if and only if at least k its arguments are set to 1 is called a *threshold function with the threshold* k. Denote by $Thr_{n,k}$ the threshold function of n variables with the threshold k.

Lemma 3.4. *For an arbitrary natural number* n, *the relation* $h(Thr_{n,\lceil n/2 \rceil}) \geq n + 1 - \sqrt{n+1}$ *holds, and for an arbitrary odd* $n \geq 3$, *the relation*

$$h(Thr_{n,(n+1)/2}) \geq n + \frac{3}{2} - \sqrt{n+2}$$

holds.

Proof. Denote $m = \lceil n/2 \rceil$. Let Γ be an optimal decision tree implementing the function $Thr_{n,m}$. Let us transform Γ as follows. We will process nonterminal nodes layer by layer starting from the root. Let v be the current node, r_v the distance from the root to v, and Γ_v the tree whose root is v. If v is assigned with the attribute x_{r_v+1}, then skip this node and proceed to the next node. Let v is assigned with an attribute x_s that differs from x_{r_v+1}. Lemma 2.3 implies that Γ is a reduced tree. Then for each path in Γ, the attributes assigned to the nonterminal nodes of this path are pairwise different. Therefore, no node in Γ_v except the root is assigned with the attribute x_s. Assign the attribute x_{r_v+1} to the node v, assign the attribute x_s to all nonterminal nodes in Γ_v which were assigned with the attribute x_{r_v+1}, and proceed to the next node.

One can see that Γ_v is a decision tree implementing the function $Thr_{n,m}(\delta_1, \ldots, \delta_{r_v}, x_{r_v+1}, \ldots, x_n)$ for some $\delta_1, \ldots, \delta_{r_v} \in \{0,1\}$. Since this function is symmetrical, the transformation keeps the function implemented by Γ_v and does not change the average depth of the tree.

Denote $\hat{\Gamma}$ the resulted tree. From the description of the transformation it follows that $\hat{\Gamma}$ is an optimal decision tree implementing the function $Thr_{n,m}$, and for $i = 1, \ldots, n$, all nodes in the i-th layer are assigned with the attribute x_i. According to Lemma 2.3, $\hat{\Gamma}$ is a reduced decision tree. One can see that for a tuple $\bar{\delta} = (\delta_1, \ldots, \delta_n) \in E_2^n$, the length of the path on which computations for $\bar{\delta}$ are performed is equal to i if and only if one of the following conditions hold:

- $\delta_i = 1$, and exactly $(m-1)$ elements of the tuple $(\delta_1, \ldots, \delta_{i-1})$ are equal to one;
- $\delta_i = 0$, and exactly $(n-m)$ elements of the tuple $(\delta_1, \ldots, \delta_{i-1})$ are equal to zero.

In other words, the length of the path is the minimum of the position of the m-th one and the position of the $(n-m+1)$-th zero in the tuple $\bar{\delta}$.

For $i = 1, \ldots, n$, there are $2^{n-i}(C_{i-1}^{m-1} + C_{i-1}^{n-m})$ tuples corresponding to the paths in $\hat{\Gamma}$ of the length i. Then the average depth of the decision tree $\hat{\Gamma}$ is equal to

$$h(\hat{\Gamma}) = 2^{-n} \sum_{i=1}^{n} i 2^{n-i}(C_{i-1}^{m-1} + C_{i-1}^{n-m}) .$$

Applying simple transformations and using the previously introduced notation $g(n, t)$, we obtain $h(\hat{\Gamma}) = h(Thr_{n,m}) = mg(n, m) + (n - m + 1)g(n, n - m + 1)$. Applying Lemma 3.2, we obtain

$$h(Thr_{n,m}) = 2(n + 1) - \frac{1}{2^n} \left(m \sum_{i=0}^{m} C_{n+1}^i + (n - m + 1) \sum_{i=0}^{n-m+1} C_{n+1}^i \right) .$$

If n is odd, then $m = (n + 1)/2$. Taking into account that $(n + 1)/2 = n - (n + 1)/2 + 1$, and

$$\sum_{i=0}^{(n+1)/2} C_{n+1}^i = 2^n + \frac{1}{2} C_{n+1}^{(n+1)/2}$$

as the number of binary tuples of the length $(n + 1)$ containing at most $(n + 1)/2$ ones, we have

$$h(Thr_{n,m}) = (n + 1) \left(1 - \frac{1}{2^{n+1}} C_{n+1}^{(n+1)/2} \right) .$$

Using a known bound from [29] (see Chap. 8, Exercise 8.5.2)

$$C_{2n}^n \leq \frac{4^n}{\sqrt{3n + 1}} , \tag{3.3}$$

we obtain that $h(Thr_{n,m}) \geq (n + 1)(1 - \sqrt{2}/\sqrt{3n + 5}) \geq (n + 1) - \sqrt{n + 1}$. Applying Lemma 3.3, we obtain the bound $h(Thr_{n,m}) \geq n + 3/2 - \sqrt{n + 2}$ for any odd $n \geq 3$.

Let n be even. Then $m = n/2$. Taking into account that $\sum_{i=0}^{n/2} C_{n+1}^i = 2^n$ and $\sum_{i=0}^{n/2+1} C_{n+1}^i = 2^n + C_{n+1}^{n/2+1}$, we have

$$h(Thr_{n,m}) = (n + 1) - \frac{1}{2^n} \left(\frac{n}{2} + 1 \right) C_{n+1}^{n/2+1} = (n + 1) \left(1 - \frac{1}{2^n} C_n^{n/2} \right) .$$

The inequality (3.3) implies $h(Thr_{n,m}) \geq (n + 1)(1 - \sqrt{2}/\sqrt{3n + 2}) \geq (n + 1) - \sqrt{n + 1}$.   □

Let $z = (\nu, f_1, \ldots, f_n)$ be a problem over a 2-valued information system. A set of terminal separable subtables $\{I_1, \ldots, I_k\}$ of the table T_z is called *compatible* if for some natural number $l \leq n$, there exist numbers $a_1, \ldots, a_l \in \{1, \ldots, n\}$, and for $i = 1, \ldots, k$, there exist tuples $\bar{\delta}_i = (\delta_1^i, \ldots, \delta_l^i) \in E_2^l$, $\bar{\delta}_i \neq \bar{\delta}_j$ for $i \neq j$, such that $I_i = T_z(f_{a_1}, \delta_1^i) \ldots (f_{a_l}, \delta_l^i)$. We will say that the terminal separable subtables $I_1, \ldots, I_k$ form a *partition of the table* T_z if $\bigcup_{i=1}^k I_i = T_z$, and $I_i \cap I_j = \emptyset$ for $i \neq j$.

Lemma 3.5. *Let* $z = (\nu, f_1, \ldots, f_n)$ *be a problem over 2-valued information system such that* $T_z = E_2^n$, *and* P *be a probability distribution for the problem* z. *Then the following statements are valid:*

a) for an arbitrary compatible set of terminal separable subtables $\{I_1, \ldots, I_k\}$ *of the table* T_z, *the inequality*

$$h(z, P) \leq n - \frac{1}{N(T_z, P)} \sum_{i=1}^k \log_2 D(I_i) N(I_i, P) \tag{3.4}$$

holds;

b) there exists a partition of the table T_z *for which the relation (3.4) holds as equality.*

Proof. Let us prove part (a) of the lemma. Let $\nu(x) \equiv \text{const} = \nu_i$ on the set of rows of the table I_i for $i = 1, \ldots, k$. Without loss of generality, assume that $I_i = T_z(f_1, \delta_1^i) \ldots (f_l, \delta_l^i)$ for some $\delta_1^i, \ldots, \delta_l^i \in E_2$, $i = 1, \ldots, k$. Let us build a decision tree Γ for the problem z in the following way.

Step 0. Build a complete binary tree of the length $(l + 1)$. For $i = 1, \ldots, l$, assign the attribute f_i to each node in the i-th layer and proceed to the first step.

Let $t \geq 0$ steps have been already performed.

Step $(t + 1)$. If each terminal node in the tree Γ has been already labeled with a number, the algorithm finishes. Otherwise, choose in Γ an unlabeled terminal node v. Denote by ξ the path from the root to the node v. If $T_z \pi(\xi) = I_i$ for some $i \in \{1, \ldots, k\}$, then label the node v with the number ν_i and proceed to the next step. Otherwise, replace the node v with a complete binary tree Γ_v of the depth $(n - l + 1)$, and for $i = 1, \ldots, n - l$, assign the attribute f_{l+i} to all nonterminal nodes in the i-th layer of the tree Γ_v. Then assign to each terminal node w of the tree Γ_v the natural number a_w defined as follows. Denote by ϕ the path from the root of the tree Γ to the node w. Since each of the attributes $f_1, \ldots, f_n$ is assigned to a node in the path ϕ and $T_z = E_2^n$, the subtable $T_z \pi(\phi)$ consists of a single row. Denote that row $\bar{\delta}$ and assume $a_w = \nu(\bar{\delta})$. Proceed to the step $(t + 2)$.

One can see that the algorithm finishes after the (2^{l+1})-th step and the resulted decision tree solves the problem z. For an arbitrary tuple $\bar{\delta} \in T_z$, denote by $\xi(\bar{\delta})$ the complete path on which computations for $\bar{\delta}$ are performed.

From the description of tree building procedure, the length of $\xi(\bar{\delta})$ is equal to l if $\bar{\delta} \in I_i$ for some $i \in \{1, \ldots, k\}$, and is equal to n otherwise. Denote $I = I_1 \cup \ldots \cup I_k$. Then the expression for the average depth of the decision tree Γ has the following form:

$$h(\Gamma, P) = \frac{1}{N(T_z, P)} \left(n \sum_{\bar{\delta} \in T_z \setminus I} P(\bar{\delta}) + l \sum_{\bar{\delta} \in I} P(\bar{\delta}) \right)$$

$$= n - \frac{1}{N(T_z, P)} \sum_{i=1}^{k} (n - l) N(I_i, P) .$$

(3.5)

Since $T_z = E_2^n$, each subtable I_i contains exactly 2^{n-l} rows, and $\log_2 D(I_i) = n - l$. Taking into account the obvious inequality $h(z, P) \leq h(\Gamma, P)$, the inequality (3.5) can be easily transformed into (3.4).

Let us prove part (b) of the lemma. Let Γ be an optimal decision tree for the problem z. Denote by $W(\Gamma)$ the set of nonterminal nodes in the decision tree Γ. For an arbitrary terminal node $w \in W(\Gamma)$, denote by $path(w)$ the path from the root to the node w. Since the tree Γ solves z, $T_z \pi(path(w))$ is a terminal subtable, $T_z \pi(path(w_1)) \cap T_z \pi(path(w_2)) = \emptyset$ for any two different nodes w_1 and w_2, and $\bigcup_{w \in W(\Gamma)} T_z(\pi(path(w))) = T_z$. For an arbitrary terminal node $w \in W(\Gamma)$, denote $I_w = T_z \pi(path(w))$ and choose the set $\{I_w : w \in W(\Gamma)\}$ as the desired partition. From Lemma 2.3 it follows that Γ is a reduced tree. Then for an arbitrary terminal node $w \in W(\Gamma)$, the nonterminal nodes of the path $path(w)$ are assigned with pairwise different attributes. From the condition $T_z = E_2^n$ it follows that the number of rows in the subtable I_w is equal to $2^{n - l_{\Gamma}(path(w))}$. Therefore, the length of the path on which computations for all rows of the table I_w are performed is equal to $(n - \log_2 D(I_w))$. Finally, we transform the expression for the average depth of the decision tree Γ:

$$h(\Gamma, P) = \frac{1}{N(T_z, P)} \sum_{\bar{\delta} \in T_z} l_{\Gamma}(\xi(\bar{\delta})) P(\bar{\delta})$$

$$= \frac{1}{N(T_z, P)} \sum_{w \in W(\Gamma)} l_{\Gamma}(path(w)) N(I_w, P)$$

$$= \frac{1}{N(T_z, P)} \sum_{w \in W(\Gamma)} (n - \log_2 D(I_w)) N(I_w, P)$$

$$= n - \frac{1}{N(T_z, P)} \sum_{w \in W(\Gamma)} \log_2 D(I_w) N(I_w, P) . \qquad \square$$

Lemma 3.6. *For an arbitrary natural number n, the minimum average depth of decision trees implementing Boolean functions $x_1 \oplus \ldots \oplus x_n$ and $x_1 \oplus \ldots \oplus x_n \oplus 1$ is equal to n.*

Proof. Let $z = (f, x_1, \ldots, x_n)$ be a problem corresponding to the function $f(x_1, \ldots, x_n) = x_1 \oplus \ldots \oplus x_n$. Let us show that T_z does not have terminal separable subtables that contain more than one row. Assume the contrary. Let there exist a word $\alpha \in \Omega_z$ such that $T_z \alpha$ is a terminal separable subtable containing at least two rows $\bar{\delta} = (\delta_1, \ldots, \delta_n)$ and $\bar{\sigma} = (\sigma_1, \ldots, \sigma_n)$, $\bar{\delta} \neq \bar{\sigma}$. Let $\delta_i \neq \sigma_i$ for some $i \in \{1, \ldots, n\}$. Then the word α does not contain the letters $(x_i, 0)$ and $(x_i, 1)$. Since $T_z = E_2^n$, the subtable $T_z \alpha$ also contains the row $\bar{\delta}^* = (\delta_1, \ldots, \delta_i - 1, \neg \delta_i, \delta_{i+1}, \ldots, \delta_n)$. Since $f(\bar{\delta}) \neq f(\bar{\delta}^*)$, we obtain a contradiction with the assumption that $T_z \alpha$ is a terminal subtable. Therefore, all terminal separable subtables of T_z consist of a single row. From part (b) of Lemma 3.5 it follows that $h(f) = n$. The lemma is proved analogously for the function $x_1 \oplus \ldots \oplus x_n \oplus 1$. □

Lemma 3.7. *Let $f(x_1, \ldots, x_n)$ be an arbitrary non-constant Boolean function. Denote by $\hat{f}$ one of the functions $f(x_1, \ldots, x_n) \wedge x_{n+1}$, $f(x_1, \ldots, x_n) \wedge \neg x_{n+1}$, $f(x_1, \ldots, x_n) \vee x_{n+1}$, $f(x_1, \ldots, x_n) \vee \neg x_{n+1}$. Then the relation $h(\hat{f}) = 1 + h(f)/2$ holds.*

Proof. Let us prove the lemma by induction on the number of essential variables of the function f. Obviously, each function that have a single essential variable can be represented (up to inessential variables) in the form $f(x) = x$ or $f(x) = \neg x$, and the lemma is valid for these functions.

Let the lemma be valid for all functions with at most $(t - 1)$ essential variables for some $t > 1$. Let f be a function with t essential variables. Denote $\hat{f} = f(x_1, \ldots, x_n) \wedge x_{n+1}$.

Let us build a decision tree $\hat{\Gamma}$ in the following way. The root of $\hat{\Gamma}$ is assigned with the attribute x_{n+1}. Two edges leave the root labeled with the numbers 0 and 1. The edge labeled with 0 enters a terminal node which is labeled with the number 0. The edge labeled with 1 enters the root of an optimal decision tree implementing the function f. The decision tree $\hat{\Gamma}$ does not contain any other nodes and edges. It is easy to see that $\hat{\Gamma}$ implements the function $\hat{f}$. According to Lemma 3.1,

$$h(\hat{\Gamma}) = 1 + \frac{h(f)}{2} . \tag{3.6}$$

To prove the lemma it is sufficient to show that $\hat{\Gamma}$ is an optimal decision tree. Assume the contrary. In this case, there exists an optimal decision tree

Γ whose root is assigned with an attribute other than x_{n+1}. Assume without loss of generality that it is the attribute x_1. For $\delta = 0, 1$, denote e_δ the edge that leaves the root of Γ and is labeled with the number δ, and denote Γ_δ the decision tree whose root the edge e_δ enters. One can see that the decision tree Γ_δ implements the function $f(\delta, x_2, \ldots, x_n) \wedge x_{n+1}$. Since f has t essential variables, at least one of the functions $f(0, x_2, \ldots, x_n)$, $f(1, x_2, \ldots, x_n)$ is a non-constant function. Let both functions possess this condition. Then the induction base implies that for $\delta = 0, 1$, the relation

$$h(f(\delta, x_2, \ldots, x_n) \wedge x_{n+1}) = 1 + \frac{h(f(\delta, x_2, \ldots, x_n))}{2} \tag{3.7}$$

holds. From Lemma 3.1 it follows that

$$h(\Gamma) = 1 + \frac{1}{2}\left(1 + \frac{h(f(0, x_2, \ldots, x_n))}{2} + 1 + \frac{h(f(1, x_2, \ldots, x_n))}{2}\right).$$

Then

$$h(\Gamma) = 2 + \frac{h(f(0, x_2, \ldots, x_n)) + h(f(1, x_2, \ldots, x_n))}{4}. \tag{3.8}$$

Let us build a decision tree G as follows. The root of G is assigned with the attribute x_1. Two edges leave the root labeled with the numbers 0 and 1. For $\delta = 0, 1$, the edge labeled with the number δ enters the root of an optimal decision tree for the function $f(\delta, x_2, \ldots, x_n)$. The tree G does not contain any other nodes and edges. One can see that G implements the function f. According to Lemma 3.1,

$$h(G) = 1 + \frac{h(f(0, x_2, \ldots, x_n)) + h(f(1, x_2, \ldots, x_n))}{2}. \tag{3.9}$$

Taking into account the inequality $h(f) \leq h(G)$ and substituting (3.9) into (3.8), we obtain that $h(\Gamma) \geq 3/2 + h(f)/2$. Comparing the last relation to (3.6), we have $h(\Gamma) > h(\hat{\Gamma})$ that contradicts the assumption that Γ is an optimal decision tree. Therefore, only one of the functions $f(0, x_2, \ldots, x_n)$, $f(1, x_2, \ldots, x_n)$ is non-constant. Suppose for the definiteness that $f(1, x_2, \ldots, x_n) \equiv \text{const}$. Then f can be represented in the form $f = x_1 \vee f(0, x_2, \ldots, x_n)$ or $f = \neg x_1 \wedge f(0, x_2, \ldots, x_n)$. The induction base implies that $h(f) = 1 + h(f(0, x_2, \ldots, x_n))/2$. The function $f(0, x_2, \ldots, x_n) \wedge x_{n+1}$ is non-constant, and for $\delta = 0$, the relation (3.7) holds. Then the relation $h(f(0, x_2, \ldots, x_n) \wedge x_{n+1}) = h(f)$ holds and implies

$$h(\hat{f}) = h(\Gamma) = 1 + \frac{h(f)}{2} + \frac{h(f(1, x_2, \ldots, x_n) \wedge x_{n+1})}{2}.$$

Comparing this relation with (3.6), we have $h(\hat{f}) \leq h(\hat{\Gamma})$ that contradicts the assumption that the tree $\hat{\Gamma}$ is not optimal. Consequently, the tree $\hat{\Gamma}$ is optimal and $h(f) = 1 + h(f)/2$. The induction step is proved analogously for other types of the function $\hat{f}$ listed in the lemma. □

For arbitrary numbers $x, \sigma \in E_2$, denote

$$
x^\sigma = \begin{cases} x, & \text{if } \sigma = 0, \\ \neg x, & \text{if } \sigma = 1. \end{cases}
$$

Proof of Proposition 3.1. For an arbitrary natural number n and a number $\sigma \in E_2$, set into correspondence to the function $f(x_1, \ldots, x_n) \equiv \sigma$ a decision tree $\Gamma_0(\sigma)$ that consists of a single terminal node labeled with σ. One can see that the tree $\Gamma_0(\sigma)$ implements f and $h(\Gamma_0(\sigma)) = 0$. Then $\mathcal{H}_B(n) = 0$. □

Proof of Proposition 3.2. For an arbitrary natural number n, a natural number $i \leq n$, and a number $\sigma \in E_2$ set into correspondence to the function $f(x_1, \ldots, x_n) = x_i^\sigma$ a decision tree $\Gamma_1(i, \sigma)$. The decision tree $\Gamma_1(i, \sigma)$ consists of one nonterminal node v labeled with the attribute x_i and two terminal nodes w_0 and w_1 labeled with the numbers 0^σ and 1^σ respectively. For $\delta = 0, 1$, there is an edge leaving v and entering w_δ, and this edge is labeled with the number δ. The tree $\Gamma_1(i, \sigma)$ does not contain other nodes and edges.

One can see that the tree $\Gamma_1(i, \sigma)$ implements f and $h(\Gamma_1(i, \sigma)) = 1$. On the other hand, a decision tree that implements a non-constant Boolean function must have at least two terminal nodes and, consequently, at least one nonterminal node. Therefore, $\Gamma_1(i, \sigma)$ is an optimal decision tree and $\mathcal{H}_B(n) = 1$. □

Proof of Proposition 3.3. Any function of n arguments from the set $S_1 \cup S_3 \cup S_5 \cup S_6 \cup P_1 \cup P_3 \cup P_5 \cup P_6$ up to argument names can be represented in the form $f_0(x_1, \ldots, x_n) = 0$, $f_1(x_1, \ldots, x_n) = 1$, $f_t^1(x_1, \ldots, x_n) = x_1 \vee \ldots \vee x_t$ or $f_t^2(x_1, \ldots, x_n) = x_1 \wedge \ldots \wedge x_t$ where $t \leq n$. Let us prove by induction on t that the relation $h(f_t^1) = h(f_t^2) = 2 - 1/2^{t-1}$ holds for $t = 1, \ldots, n$. If $t = 1$, then $f_t^1 \equiv f_t^2 \equiv x_1$ and $h(f_t^1) = h(f_t^2) = 1$. Let the relation be valid for each i less than t. From Lemma 3.7 it follows that $h(f_t^1) = 1 + h(f_{t-1}^1)/2$ and $h(f_t^2) = 1 + h(f_{t-1}^2)/2$. According to the inductive hypothesis, $h(f_{t-1}^1) = h(f_{t-1}^2) = 2 - 1/2^{t-2}$. Then $h(f_t^1) = h(f_t^2) = 1 + (2 - 1/2^{t-2})/2 = 2 - 1/2^{t-1}$. One can see that $\max_{t \in \{1, \ldots, n\}} h(f_t^1) = \max_{t \in \{1, \ldots, n\}} h(f_t^2) = 2 - 2^{1-n}$ and the maximum is reached on the functions f_n^1 and f_n^2. Then validity of the proposition follows from the fact that for $n \geq 1$, each of the classes $S_1, S_3, S_5, S_6, P_1, P_3, P_5, P_6$ contains at least one of the functions f_n^1 and f_n^2. □

Proof of Proposition 3.4. Each of the closed classes listed in the proposition contains at least one of the functions $x_1 \oplus \ldots \oplus x_n$ and $x_1 \oplus \ldots \oplus x_n \oplus 1$. Using Lemma 3.6, we obtain that $\mathcal{H}_B(n) \geq n$. On the other hand, obviously, any Boolean function of n arguments can be implemented by a decision tree of the average depth at most n. Consequently, $\mathcal{H}_B(n) = n$.   □

Proof of Proposition 3.5. For $n = 2k + 1$, both classes contain the function $x_1 \oplus \ldots \oplus x_n$ and validity of the proposition is proved analogously to Proposition 3.4. For $n = 2k$, none of the classes contain a function with n essential variables and both classes contain the function $x_1 \oplus \ldots \oplus x_{n-1}$. Therefore, $\mathcal{H}_B(n) = n - 1$.   □

Proof of Proposition 3.6. For $n = 2k + 1$, the class contains the function $x_1 \oplus \ldots \oplus x_n$, and validity of the proposition is proved analogously to Proposition 3.4. Let $n = 2k$, and f be the function such that $h(f) = \mathcal{H}_B(n)$. Let $z = (f, x_1, \ldots, x_n)$ be a problem corresponding to the function f. Consider a sequence of rows $\bar{\delta}^0, \ldots, \bar{\delta}^n \in T_z$ where $\bar{\delta}^0 = (0, \ldots, 0, 0)$, $\bar{\delta}^1 = (0, \ldots, 0, 1)$, $\bar{\delta}^2 = (0, \ldots, 1, 1)$, ..., $\bar{\delta}^n = (1, \ldots, 1, 1)$. Note that $f(\bar{\delta}^0) = 0$ and $f(\bar{\delta}^n) = 1$ because $f \in C_4$. Since $n = 2k$, the relation $f(\bar{\delta}^i) = f(\bar{\delta}^{i+1})$ holds for some $i \in \{0, \ldots, n - 1\}$. Thus the table T_z has a terminal separable subtable $T_z(x_0, 0)(x_1, 0) \ldots (x_{i-1}, 0)(x_{i+1}, 1) \ldots (x_n, 1)$ containing exactly two rows: $\bar{\delta}^i$ and $\bar{\delta}^{i+1}$. From part (a) of Lemma 3.5 it follows that $h(f) \leq n - 2^{1-n}$. Then $\mathcal{H}_B(n) \leq n - 2^{1-n}$.

Consider the function $f(x_1, \ldots, x_n) = x_1 \wedge \ldots \wedge x_n \vee (x_1 \oplus \ldots \oplus x_n)$. Let z be the problem corresponding to the function f. One can see that the table T_z has n terminal separable subtables $I_1, \ldots, I_n$, $I_i = T_z(x_1, 1) \ldots (x_{i-1}, 1)(x_{i+1}, 1) \ldots (x_n, 1)$, containing two rows and does not have other terminal separable subtables containing more than one row. The subtables $I_1, \ldots, I_n$ have the common row $(1, \ldots, 1)$. Thus any partition of the table T_z can contain only one of the subtables $I_1, \ldots, I_n$. From part (b) of Lemma 3.5 it follows that $h(f) = n - 2^{1-n}$. Then $\mathcal{H}_B(n) = n - 2^{1-n}$.   □

Proof of Proposition 3.7. For $n = 2k + 1$, both classes contain one of the functions $x_1 \oplus \ldots \oplus x_n$, $x_1 \oplus \ldots \oplus x_n \oplus 1$ and validity of the proposition is proved analogously to Proposition 3.4. Let $n = 2k$. Since $D_1 \subseteq C_4$ and $D_3 \subseteq C_4$, validity of the upper bound follows from Proposition 3.6.

Consider a function $f(x_1, \ldots, x_n)$ defined as follows: on a tuple $\bar{\delta} = (\delta_1, \ldots, \delta_n)$ it takes the value $(\delta_1 \oplus \ldots \oplus \delta_n)$ if the number of zeros in $\bar{\delta}$ is greater than the number of ones, the value $(\delta_1 \oplus \ldots \oplus \delta_n \oplus 1)$ if the number of zeros in $\bar{\delta}$ is less than the number of ones, and the value δ_1 if the number of zeros is equal to the number of ones. Let us show that any terminal separable subtable I of the table T_z contains at most one row in which the number of zeros differs from the number of ones. Assume the contrary, i.e. the table I contains at least two such rows. Let the number of zeros in the first row exceed the number of ones. If the second row possesses the same condition, then the subtable I contains two rows $\bar{\delta}^1$ and $\bar{\delta}^2$ in which the number of zeros is greater than the number of ones and which differ exactly in one digit. If the number of ones in the second row exceeds the number of zeros, then the subtable I contains two rows $\bar{\delta}^1$ and $\bar{\delta}^2$ with $(n/2 - 1)$ and $(n/2 + 1)$ ones respectively. According to the definition of the function, $f(\bar{\delta}_1) \neq f(\bar{\delta}_2)$ for both cases that contradicts the assumption that I is a terminal subtable. The case when the first row contains more zeros than ones is considered analogously. Therefore, each terminal separable subtable of the table T_z contains at most one row in which the number of zeros differs from the number of ones. It implies that any terminal separable subtable of T_z contains at most two rows and each two-row subtable contains a row with $n/2$ zeros and $n/2$ ones. Obviously, the table T_z contains $C_n^{n/2}$ such rows. Then any partition of the table T_z contains at most $C_n^{n/2}$ subtables with two rows and does not contain subtables with a greater number of rows. According to part (b) of Lemma 3.5, the relation $h(f) \geq n - 1/2^{n-1}C_n^{n/2}$ holds. Using the bound (3.3), we obtain

$$\mathcal{H}_B(n) \geq h(f) \geq n - \frac{2\sqrt{2}}{\sqrt{3}} \frac{1}{\sqrt{n}} \geq n - \frac{1.7}{\sqrt{n}} \ . \qquad \square$$

Proof of Proposition 3.8. It is not hard to show that each class contains the function $Thr_{n,\lceil n/2 \rceil}$. Validity of the lower bound of the lemma immediately follows from Lemma 3.4. Let us prove validity of the upper bound. Let $f(x_1, \ldots, x_n)$ be a function for which the equality $h(f) = \mathcal{H}_B(n)$ holds. Denote $z = (f, x_1, \ldots, x_n)$ the problem corresponding to the function f. Consider the value of the function $f(\bar{\delta})$ on the tuple $\bar{\delta} = (\delta_1, \ldots, \delta_n)$ where $\delta_1 = \delta_2 = \ldots = \delta_m = 1$, $\delta_{m+1} = \delta_{m+2} = \ldots = \delta_n = 0$, $m = \lfloor n/2 \rfloor$. If $f(\bar{\delta}) = 1$, then f takes the value 1 on all tuples in which the first m digits are set to 1. Then the table T_z has a terminal separable subtable $T_z(x_1, 1)(x_2, 1) \ldots (x_m, 1)$, containing $2^{n-m} = 2^{\lfloor (n+1)/2 \rfloor}$ rows. From part (a) of Lemma 3.5 it follows that

$$h(f) \leq n - \lfloor (n+1)/2 \rfloor \, 2^{-\lfloor (n+1)/2 \rfloor} \ . \qquad (3.10)$$

If $f(\bar{\delta}) = 0$, then the function f takes the value 0 on all tuples in which the last $(n - m)$ digits are set to 0. Thus the table T_z has a terminal separable subtable $T_z(x_{m+1}, 0)(x_{m+2}, 0) \ldots (x_n, 0)$ containing $2^m = 2^{\lfloor n/2 \rfloor}$ rows. From part (a) of Lemma 3.5 it follows that

$$h(f) \leq n - \lfloor n/2 \rfloor \, 2^{-\lfloor n/2 \rfloor} \, . \tag{3.11}$$

By taking the weakest bound of (3.10) and (3.11), we obtain $\mathcal{H}_B(n) = h(f) \leq n - \lfloor n/2 \rfloor 2^{-\lfloor n/2 \rfloor}$. $\qquad \square$

Proof of Proposition 3.9. It is easy to show that the class D_2 contains the function $Thr_{n,(n+1)/2}$ for an arbitrary odd n, and the function $Thr_{n-1,n/2}$ for an arbitrary even n. Then validity of the lower bound for an arbitrary $n \geq 3$ follows from Lemma 3.4. Validity of the lower bound for $n = 1, 2$ can be proved by a direct check. The upper bound follows from the relation $D_2 \subset M_1$ and Proposition 3.8. $\qquad \square$

Proof of Proposition 3.10. One can see that each function of n arguments from the set $F_1^\infty \cup F_4^\infty \cup F_5^\infty \cup F_8^\infty$ can be represented in the form $f_n^0(x_1, \ldots, x_n) = x_i \vee \phi_{n-1}^0(x_1, \ldots, x_{i-1}, x_{i+1}, \ldots, x_n)$ or $f_n^1(x_1, \ldots, x_n) = x_i \wedge \phi_{n-1}^1(x_1, \ldots, x_{i-1}, x_{i+1}, \ldots, x_n)$. If ϕ_{n-1}^δ is a constant function for $\delta = 0$ or $\delta = 1$, then $h(f_n^\delta) \leq 1$. Let the function ϕ_{n-1}^δ be non-constant. According to Lemma 3.7, $h(f_n^\delta) = 1 + h(\phi_{n-1}^\delta)/2$. Since the function ϕ_{n-1}^δ has at most $(n-1)$ essential variables, $h(\phi_{n-1}^\delta) \leq n - 1$ and $h(f_n^\delta) \leq 1 + (n-1)/2$. This relation holds for all functions and, consequently, $\mathcal{H}_B(n) \leq (n+1)/2$.

For $\sigma = 0, 1$, consider the functions $f_n^{0,\sigma}(x_1, \ldots, x_n) = x_1 \vee (x_2 \oplus \ldots \oplus x_n \oplus \sigma)$ and $f_n^{1,\sigma}(x_1, \ldots, x_n) = x_1 \wedge (x_2 \oplus \ldots \oplus x_n \oplus \sigma)$. One can see that for an arbitrary natural n, each of the classes $F_1^\infty, F_4^\infty, F_5^\infty, F_8^\infty$ contains at least one of these functions. According to Lemma 3.7, $h(f_n^{0,\sigma}) = h(f_n^{1,\sigma}) = 1 + h(x_2 \oplus \ldots \oplus x_n \oplus \sigma)/2 = (n+1)/2$. Then $\mathcal{H}_B(n) = (n+1)/2$. $\qquad \square$

Proof of Proposition 3.11. One can see that each class contains one of the functions

$$f_1 = Thr_{n-1, \lceil (n-1)/2 \rceil}(x_1, \ldots, x_{n-1}) \wedge x_n \, ,$$
$$f_2 = Thr_{n-1, \lceil (n-1)/2 \rceil}(x_1, \ldots, x_{n-1}) \vee x_n$$

for every $n > 1$. According to Lemma 3.4, $h(Thr_{n-1, \lceil (n-1)/2 \rceil}) \geq n - \sqrt{n}$. According to Lemma 3.7, $h(f_1) = h(f_2) \geq 1 + (n - \sqrt{n})/2$. Validity of the lower bound for $n = 1$ can be proved by a direct check. Validity of the upper bound follows from the relations $F_2^\infty \subseteq F_1^\infty$, $F_3^\infty \subseteq F_4^\infty$, $F_6^\infty \subseteq F_5^\infty$ $F_7^\infty \subseteq F_8^\infty$ and Proposition 3.10. $\qquad \square$

Proof of Proposition 3.12. Validity of the lower bound follows from the relations $F_1^\mu \supseteq F_1^\infty$, $F_4^\mu \supseteq F_4^\infty$, $F_5^\mu \supseteq F_5^\infty$, $F_8^\mu \supseteq F_8^\infty$, $\mu \geq 2$ and Proposition 3.10.

Let $B = F_4^2$, $f(x_1, \ldots, x_n)$ be a function such that $h(f) = \mathcal{H}_B(n)$. Let $z = (f, x_1, \ldots, x_n)$ be the problem corresponding to f. Since $F_4^2 \subseteq C_2$, the relation $f(1, \ldots, 1) = 1$ holds. The lower bound implies that f differs from the constant 1. Then there exists a number k, $1 \leq k \leq n$ such that f takes the value 1 on all tuples containing less than k zeros, and there exists a tuple containing exactly k zeros on which the function takes the value 0. Then the table T_z has a terminal separable subtable $T_z(x_1, 1)(x_2, 1) \ldots (x_{n-k+1}, 1)$ that contains 2^{k-1} rows. According to part (a) of Lemma 3.5, the inequality

$$h(f) \leq n - \frac{k-1}{2^{k-1}} \tag{3.12}$$

holds. Without loss of generality, assume that the function takes the value 0 on the tuple $\bar{\delta} = (\delta_1, \ldots, \delta_n)$ in which $\delta_1 = \delta_2 = \ldots = \delta_k = 0$ and $\delta_{k+1} = \delta_{k+2} = \ldots = \delta_n = 1$. Then in each tuple $\bar{\delta} = (\delta_1, \ldots, \delta_n)$ such that $f(\bar{\delta}) = 0$, at least one of the first k digits $\delta_1, \ldots, \delta_k$ is set to zero. Therefore, the table T_z has a terminal separable subtable $T_z(x_1, 1)(x_2, 1) \ldots (x_k, 1)$ containing 2^{n-k} rows. According to part (a) of Lemma 3.5, the inequality

$$h(f) \leq n - \frac{n-k}{2^{n-k}} \tag{3.13}$$

holds.

The weakest of the bounds (3.12) and (3.13) reaches the maximum on $k = \lfloor (n+1)/2 \rfloor$. Consequently, $\mathcal{H}_B(n) \leq n - \lfloor n/2 \rfloor 2^{-\lfloor n/2 \rfloor}$. The upper bound for $B = F_8^2$ is proved analogously. The upper bound for the remaining classes follows from the relations $F_1^\mu \subseteq F_4^2$, $F_4^\mu \subseteq F_4^2$, $F_5^\mu \subseteq F_8^2$, $F_8^\mu \subseteq F_8^2$ that are valid for any natural $\mu \geq 2$. $\square$

Proof of Proposition 3.13. Validity of the lower bound follows from the relations $F_2^\mu \supseteq F_2^\infty$, $F_3^\mu \supseteq F_3^\infty$, $F_6^\mu \supseteq F_6^\infty$, $F_7^\mu \supseteq F_7^\infty$, $\mu \geq 2$ and Proposition 3.11.

Validity of the upper bound follows from the relations $F_2^\mu \subseteq F_1^\mu$, $F_3^\mu \subseteq F_4^\mu$, $F_6^\mu \subseteq F_5^\mu$ $F_7^\mu \subseteq F_8^\mu$, $\mu \geq 2$ and Proposition 3.12. $\square$

3.2   On Branching Programs with Minimum Average Depth

This section considers a possibility of joint optimization of time and space complexity. For this purpose, a decision tree is represented in a compact form named branching program. According to Theorem 3.4, the requirement to a branching program to have the minimum average weighted depth is rather strong, since all branching programs with the minimum average weighted

depth are read-once. This fact reveals a contradiction between time and space
complexity requirements, because many problems have high lower bounds on
the number of nodes in a read-once branching program. The section concludes with description of several problems, for which the number of nodes
in a branching program with the minimum average weighted depth grows
exponentially with the number of attributes.

Let $U = (A, F)$ be a 2-valued information system, and Ψ a weight function
for U. A *branching program* for the problem $z = (\nu, f_1, \ldots, f_n)$ over U is a
finite oriented acyclic graph in which:

a) at least one edge enters each node except one called *the root of the branching program*;
b) each terminal node (a node that does not have outgoing edges) is labeled
with a number from ω.
c) two edges leave each nonterminal node, labeled with the numbers 0 and
1 respectively;
d) each nonterminal node is assigned with an attribute from the set $\{f_1, \ldots, f_n\}$.

A path from the root to a terminal node is called *complete*. A branching
program is called *read-once* if in each complete path, all nonterminal nodes
are assigned with pairwise different attributes.

Let G be a branching program for the problem z. For an arbitrary complete path ξ in G, let us define the subtable $T_z\pi(\xi)$ of the table T_z and the
path weight $\Psi(\xi)$ in the same way as it is defined for decision trees. For
an arbitrary row $\bar{d} \in T_z$, denote by $\xi^{\bar{d}}$ the complete path in G such that
$\bar{d} \in T_z\pi(\xi^{\bar{d}})$. We will say that a *branching program* G *solves the problem* z
if for each row $\bar{d} \in T_z$, the terminal node of the path $\xi^{\bar{d}}$ is labeled with the
number $\nu(\bar{d})$. Let P be a probability distribution for the problem z. The value
$h_\Psi(G, P, z) = \sum_{\bar{d} \in T_z} \Psi(\xi^{\bar{d}}) P(\bar{d})/N(T_z, P)$ is called *P-average weighted depth
of the branching program* G. A branching program G for the problem z that
solves z and has the minimum P-average weighted depth is called *optimal
for Ψ, z and P*.

Theorem 3.4. *Let U be a 2-valued information system, Ψ a weight function
for U, z a problem over U, and P a probability distribution for z. Let G be a
branching program for z that solves z and is optimal for Ψ, z and P. Then
G is a read-once branching program.*

Proof. For an arbitrary node v, we will call *v-subprogram of the branching
program* G the set of nodes and edges from G to which an oriented path from
v exists. Let v be a node such that for each node w of v-subprogram, each
path from the root of G to w contains v. Let v have $k > 1$ incoming edges

$r_1, \ldots, r_k$. For $i = 2, \ldots, k$, let us add to the program G a subprogram G_i that coincide to v-subprogram and transform G so that the edge r_i enters the root of the subprogram G_i. Let us repeat this transformation until at most one edge enters each node in G. Denote the resulted graph Γ. One can see that Γ is a decision tree for the problem z that solves z and is optimal for Ψ, z and P.

Assume G is not a read-once branching program. Then there is a complete path ξ in Γ containing two nonterminal nodes v_1 and v_2 which are assigned with the same attribute f. Let v_1 precede v_2 in the path ξ. Denote e the edge that leaves v_1 and is contained in the path ξ, and σ the number assigned to e. Denote ξ_2 the path from the root of Γ to the node v_2. One can see that either $T_z \pi(\xi_2) = \emptyset$ or $T_z \pi(\xi_2)(f, \delta) = T_z \pi(\xi_2)$ for some $\delta \neq \sigma$. Then the node v_2 is not essential and the tree Γ is not reduced. According to Lemma 2.3, the tree Γ is not optimal for Ψ, z and P. Then the branching program G is not optimal for Ψ, z and P which contradicts the premise of the theorem and thus concludes the proof. $\square$

Let us conclude with some examples of problems for which the minimum number of nodes in the branching program with the minimum average weighted depth grows exponentially with the number of attributes. For an arbitrary Boolean function $f(x_1, \ldots, x_n)$, we will say that a branching program *implements* f if it solves the problem $z = (f, x_1, \ldots, x_n)$.

In [66], it is shown that a read-once branching program implementing the function $Mult : \{0,1\}^{2n} \to \{0,1\}$ (the middle bit in the product of two n-bit integers) contains at least $2^{\Omega(\sqrt{n})}$ nodes. In [83, 84, 85], a function $n/2 - Clique - Only : \{0,1\}^{n^2} \to \{0,1\}$ is considered that takes as input an incidence matrix for a graph with n nodes. The function takes the value 1 if and only if the graph contains a $n/2$-clique and does not contain other edges. It is shown that a read-once branching program implementing the function $n/2 - Clique - Only$ contains at least $2^{\Omega(n)}$ nodes. Note that there is a branching program implementing $n/2 - Clique - Only$ such that it has $O(n^3)$ nodes, and any attribute appears at most twice in each complete path. In [59], it is shown that a read-once branching program implementing the characteristic function of Bose-Chaudhuri codes contains at least $exp(\Omega(\sqrt{n}/2))$ nodes. Theorem 3.4 shows that the branching programs that are optimal relative to the average weighted depth have the same or greater number of nodes than the read-once branching programs with the minimum number of nodes.

Chapter 4
Algorithms for Decision Tree Construction

The study of algorithms for decision tree construction was initiated in 1960s. The first algorithms are based on the separation heuristic [13, 31] that at each step tries dividing the set of objects as evenly as possible. Later Garey and Graham [28] showed that such algorithm may construct decision trees whose average depth is arbitrarily far from the minimum. Hyafil and Rivest in [35] proved NP-hardness of DT problem that is constructing a tree with the minimum average depth for a diagnostic problem over 2-valued information system and uniform probability distribution. Cox et al. in [22] showed that for a two-class problem over information system, even finding the root node attribute for an optimal tree is an NP-hard problem.

Several exact algorithms of decision tree construction are known but, as could be expected, none of them have polynomial time complexity in general case. The algorithms based on dynamic programming [27, 60, 76] build decision tree bottom-up by synthesizing a tree for a table from trees for its separable subtables. The algorithms based on branch-and-bound technique perform depth-first search in the space of possible tree prefixes [9, 73]. The second method is more complex from the computational point of view, but it can serve as a base for approximation algorithms that use heuristics to guide search. A combination of the two approaches is described in [42]. There are also algorithms that use logic methods to analyze the function being implemented like finding function implicants [11] or T-terms [82]. A comprehensive survey of the algorithms can be found in [44].

Most of approximate algorithms for decision tree construction are greedy. These algorithms construct trees in a top-down fashion by minimizing some data impurity function at each step. Activity of a variable [43], entropy [70, 78] and Gini index [8] are widely used as data impurity functions. For some problems, a detailed analysis of existence of algorithms with a guaranteed approximation ratio has been performed. Adler and Heeringa [32]

I. Chikalov: Average Time Complexity of Decision Trees, ISRL 21, pp. 61–78.
springerlink.com   © Springer-Verlag Berlin Heidelberg 2011

proved absence of polynomial-time approximation scheme for DT problem unless $P = NP$ and described an algorithm that has $(\ln n + 1)$ approximation ratio. Chakaravarthy et al. [12] generalized the results to k-DT that is construction of a decision tree with the minimum average depth for a diagnostic problem over k-valued information system and an arbitrary probability distribution. They proved NP-hardness of $\Omega(\log n)$ approximation and described an algorithm that has $O(\log k \log n)$ approximation ratio. A similar problem, building a tree with the minimum average depth for a binary classification problem over a 2-valued information system and uniform probability distribution, is surprisingly harder. In [32], an approximation-preserving reduction of the problem to $ConDT$ is done that is building the minimum size tree for a binary classification problem over 2-valued information system. For the latter problem, Alekhnovich et al. [3] proved absence of polynomial time $c \ln n$-approximation for any constant c unless $NP \subseteq DTIME[2^{m^\epsilon}]$ for some $\epsilon < 1$.

The chapter is devoted to theoretical and experimental study of several exact and approximate algorithms for decision tree construction. It consists of four sections. The first section describes an algorithm $\mathcal{A}$ based on dynamic programming. The idea is close to [42], but it was devised by the author independently in collaboration with Dr. Moshkov. The algorithm takes as input a decision table and finds the set of all so-called irredundant decision trees that have the minimum average weighted depth. The second section experimentally estimates the approximation ratio of several greedy algorithms on data sets from UCI Machine Learning Repository [25]. The third section describes using $\mathcal{A}$ for calculating exact values of the Shannon type function $H(n)$ for the class of monotone Boolean functions for small n. The fourth section contains experimental results of applying $\mathcal{A}$ for building an optimal tree for corner point detection [74], a technique used in computer vision to track objects.

Some results of this chapter have been published in [20, 21, 55].

4.1   Algorithm $\mathcal{A}$ for Decision Tree Construction

In this section, an algorithm is considered that builds an optimal decision tree with the minimum average weighted depth for a problem represented in the form of decision table. The idea of the algorithm is based on dynamic programming [27, 42, 60, 76].

4.1.1   Representation of Set of Irredundant Decision Trees

Let $U = (A, F)$ be an information system, and $z = (\nu, f_1, \dots, f_n)$ a problem over U. Let T be a separable subtable of T_z. For $i \in \{1, \dots, n\}$, denote $E(T, i)$ the set of numbers contained in the i-th column of the table T, and denote $E(T) = \{i : i \in \{1, \dots, n\}, |E(T, i)| \geq 2\}$.

Among decision trees for the problem z that solve z we distinguish *irredundant* decision trees. Consider an arbitrary node w of the tree Γ and its corresponding separable subtable $T = T_z \pi(path(\Gamma, w))$. Let T be a terminal subtable, and $\nu(x) \equiv r$ on the set of rows of the table T for some $r \in \omega$. Then w is a terminal node labeled with r. Let T be a nonterminal subtable. Then w is labeled with an attribute f_i where $i \in E(T)$. Finally, each node w such that $T_z \pi(path(\Gamma, w)) = \emptyset$ is labeled with the number 0.

The following proposition shows that among irredundant decision trees, at least one has the minimum average weighted depth.

Proposition 4.1. *Let U be an information system, Ψ a weight function for U, z a problem over U, and P a probability distribution for z. Then there exists an irredundant decision tree that is optimal for Ψ, z and P.*

Proof. Let Γ be a decision tree for the problem z that solves z, and Γ be optimal for Ψ, z and P. Let us consider an algorithm that transforms Γ into an irredundant decision tree. The algorithm sequentially processes all nodes of the tree Γ. Let w be the current node. Denote $T = T_z \pi(path(w))$. The algorithm tries to apply the following rules to each node.

- If $T = \emptyset$, then replace the subtree whose root is w with a single node labeled with 0;

- If T is a terminal subtable and $\nu(x) \equiv r$ on the set of rows of the table T, then replace the subtree whose root is w with a single node labeled with r;

- Let T be a nonterminal subtable and w be labeled with an attribute f_i, $i \notin E(T)$. Then $E(T, i) = \{\delta\}$ for some $\delta \in E_k$. Denote by Γ_δ the decision tree whose root the edge leaving w and labeled with δ enters. Then replace the subtree whose root is w with Γ_δ.

Since each node is considered at most once, the algorithm ends in a finite number of steps. Denote the resulted decision tree by $\hat{\Gamma}$. One can see that $\hat{\Gamma}$ is an irredundant decision tree for T. Obviously, the applied transformation does not increase the complexity and thus $\hat{\Gamma}$ remains optimal.   □

Let T be a separable subtable of T_z. Let us define a problem z_T corresponding to the table T. Let $T_z = \{\bar{d}_1, \ldots, \bar{d}_s\}$ and $Q_1, \ldots, Q_s \in A$ be equivalence classes such that $(f_1(q_i), \ldots, f_n(q_i)) = \bar{d}_i$ for any $q_i \in Q_i$. Let $T = \{\bar{d}_{i_1}, \ldots, \bar{d}_{i_t}\}$. Then z_T is the problem with the same description as z over the information system $(Q_{i_1} \cup \ldots \cup Q_{i_t}, F)$. Note that z_{T_z} is the initial problem.

Denote by $Tree^*(T)$ the set of irredundant decision trees for the problem z_T. Assume technically that for $T = \emptyset$, the set $Tree^*(T)$ contains a single tree that is a node labeled with the number 0. Consider an algorithm $\mathcal{B}$ for construction of the graph $\Delta(z)$, which represents in some sense the set $Tree^*(T_z)$. Nodes of this graph are some separable subtables of the table T_z. During each step, the algorithm processes exactly one node and marks this node with the symbol *. The algorithm starts with the graph which consists of one node T_z, and finishes when all nodes of the graph are processed.

Let the algorithm have performed p steps. Describe the step $(p+1)$. If in the considered graph all nodes have already been processed, then the algorithm finishes, and the considered graph is $\Delta(z)$. Let the graph have an unprocessed node (table) T. If T is a terminal subtable and $\nu(x) \equiv r$ on the set of rows of the table T, then label the considered node with the number r, mark it with the symbol * and pass to the step $(p+2)$.

Let T be a nonterminal subtable. For each $i \in E(T)$, draw from the node T a bundle of edges. Let $E(T, i) = \{\delta_1, \ldots, \delta_t\}$. Then draw t edges from T, and label these edges with the pairs $(f_i, \delta_1), \ldots, (f_i, \delta_t)$ respectively. These edges enter the nodes $T(f_i, \delta_1), \ldots, T(f_i, \delta_t)$. If some of these nodes are not in the graph, then add these nodes to the graph. The algorithm marks the node T with the symbol * and proceeds to the step $(p+2)$.

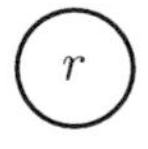

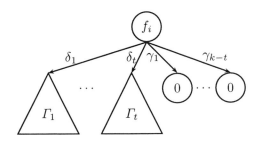

Fig. 4.1 Trivial decision tree

Fig. 4.2 Aggregated decision tree

Now for each node of the graph $\Delta(z)$, we describe the set of decision trees corresponding to it. It is clear that $\Delta(z)$ is a directed acyclic graph. A graph node is called *terminal* if it does not have outgoing edges. We will "move" from terminal nodes which are labeled with numbers to the node T_z. Let T be a node which is labeled with the number r. Then the only trivial decision tree depicted on Fig. 4.1 corresponds to the considered node. Let T be a node (table), such that $\nu(x) \not\equiv \text{const}$ on the set of rows of T. Let $i \in E(T)$, $E(T, i) = \{\delta_1, \ldots, \delta_t\}$, and $E_k \setminus E(T, i) = \{\gamma_1, \ldots, \gamma_{k-t}\}$. Let $\Gamma_1, \ldots, \Gamma_t$ be decision trees from the sets corresponding to the nodes $T(f_i, \delta_1), \ldots, T(f_i, \delta_t)$. Then the decision tree depicted on Fig. 4.2 belongs to the set of decision trees, which corresponds to the node T. All such decision trees belong to the considered set. This set does not contain any other decision trees.

For any node T, denote by $Tree(T)$ the set of decision trees corresponding to T described by the graph $\Delta(z)$. The following proposition shows that $\Delta(z)$ represents all irredundant decision trees for the problem z.

Proposition 4.2. *Let U be an information system, and z a problem over U. Let T be a node in the graph $\Delta(z)$. Then $Tree(T) = Tree^*(T)$.*

Proof. Prove the proposition by induction on the nodes of the graph $\Delta(z)$. For each terminal node T, there is only one irredundant decision tree depicted on Fig. 4.1 and the set $Tree(T)$ contains only this tree. Let T be a nonterminal node and the proposition hold for all its descendants. Consider an arbitrary decision tree $\Gamma \in Tree(T)$. Obviously, Γ contains more than one node. Let the root of Γ be labeled with an attribute f_i. For each $\delta \in E_k$, denote by Γ_δ the decision tree connected to the root of Γ with the edge labeled with the number δ. From the definition of the set $Tree(T)$ it follows that i is contained in the set $E(T)$; for each $\delta \in E(T, i)$, the decision tree Γ_δ belongs to the set $Tree(T(f_i, \delta))$; and for each $\delta \notin E(T, i)$, the decision tree Γ_δ is a single node labeled with the number 0. According to the induction base, the tree Γ_δ is an irredundant decision tree for the problem $z_{T(f_i, \delta)}$. Then the tree Γ is an irredundant decision tree for the table z_T, so $Tree(T) \subseteq Tree^*(T)$.

Now consider an arbitrary irredundant decision tree $\hat{\Gamma}$ for the problem z_T. According to the definition of irredundant tree, the root of $\hat{\Gamma}$ is labeled with an attribute f_i, $i \in E(T)$, and the subtrees whose roots are nodes in the second floor are irredundant decision trees for the corresponding descendants of the node T. Then according to the definition of the set $Tree(T)$, the tree $\hat{\Gamma}$ belongs to $Tree(T)$, and $Tree(T) = Tree^*(T)$.   □

The following proposition gives upper and lower bounds on the time complexity of the algorithm $\mathcal{B}$ (further we assume that k is fixed and do not study dependence of the algorithm time complexity on k).

Proposition 4.3. *For an arbitrary problem* $z = (\nu, f_1, \ldots, f_n)$ *represented in the form of decision table* T_z^*, *the working time of the algorithm* $\mathcal{B}$ *is proportional to the number of rows* $D(T_z)$ *if* T_z *is a terminal table. If* T_z *is a nonterminal table, the working time of the algorithm* $\mathcal{B}$ *is bounded from below by the maximum of the values* n, $|S(z)|$, $D(T_z)$ *and is bounded from above by a polynomial on these values.*

Proof. At start, the algorithm $\mathcal{B}$ reads $D(T_z)$ values of $\nu(x)$ function to check if T_z is a terminal table. If the table T_z is terminal, the algorithm builds the graph $\Delta(z)$ that consist of a single node and finishes, so the statement obviously hold.

Let T_z^* be a nonterminal table. From the definition of a problem over information system it follows that $|E(T_z)| = n$, so the algorithm builds n bundles of edges leaving the root. One can see that the algorithm $\mathcal{B}$ performs at least $|S(z)|$ steps, so the lower bound holds.

The number of steps of the algorithm $\mathcal{B}$ is limited from above by the number of nonterminal subtables and their immediate descendants in the graph $\Delta(z)$ that is at most $|S(z)|nk$. For a table T, construction of the sets $E(T)$ and $E(T, i)$ takes a linear time on the length of the table representation, i.e. $nD(T)$. It is easy to implement a procedure which given a subtable T checks if the corresponding node presents in the graph, and has a polynomial time complexity on $|S(z)|$ and $D(T)$. While processing a nonterminal table T, the algorithm needs to build a set of subtables of the form $T(f_i, \sigma)$ that can be done in a polynomial time on n and $D(T)$.

Then the total working time of the algorithm is bounded from above by a polynomial on $|S(z)|$, n and $D(T)$.   □

4.1.2 Procedure of Optimization

Let us describe a procedure which transforms the graph $\Delta(z)$ into its proper subgraph $\Delta_{\Psi,P}(z)$. It begins from the terminal nodes and moves to the node T_z. The procedure assigns a number to each node and possibly removes some bundles of edges which start in the considered node. First, the number 0 is assigned to each terminal node. Consider a node T which is not terminal and a bundle of edges which starts in this node. Let the edges be labeled with pairs $(f_i, \delta_1), \ldots, (f_i, \delta_t)$, and they enter the nodes $T(f_i, \delta_1), \ldots, T(f_i, \delta_t)$ to which numbers $p_1, \ldots, p_t$ have been already assigned. Then assign the number $\Psi(f_i)N(T, P) + \sum_{j=1}^{t} p_j$ to the considered bundle.

Let p be the minimum of the numbers assigned to the bundles starting in T. The procedure assigns p to the node T and removes the bundles starting in T which are assigned with numbers greater than p. After all nodes are processed, the procedure removes from the graph all nodes such that there is no directed path from the node T_z to the considered node. Denote the resulted graph by $\Delta_{\Psi,P}(z)$.

As each nonterminal node keeps at least one bundle of the outgoing edges, all terminal nodes of $\Delta_{\Psi,P}(z)$ are terminal nodes of the graph $\Delta(z)$. As it was described earlier, we can set to correspondence a set of decision trees $Tree_{\Psi,P}(T)$ to each node T of $\Delta_{\Psi,P}(z)$. One can see that all these decision trees belong to the set $Tree(T)$. Denote by $Tree_{\Psi,P}^*(T)$ the subset of $Tree(T)$ containing all decision trees that are optimal relative to the average weighted depth, i.e. $Tree_{\Psi,P}^*(T) = \{\hat{\Gamma} \in Tree(T) : h_\Psi(\hat{\Gamma}, z_T, P) = \min_{\Gamma \in Tree(T)} h_\Psi(\Gamma, z_T, P)\}$. The following theorem shows that the optimization procedure removes all and only non-optimal decision trees.

Theorem 4.1. *Let U be an information system, Ψ a weight function, z a problem over U, and P a probability distribution for z. Let T be an arbitrary node in the graph $\Delta(z)$. Then $Tree_{\Psi,P}(T) = Tree_{\Psi,P}^*(T)$.*

We preface proof of the theorem by the following lemma.

Lemma 4.1. *Let U be an information system, Ψ a weight function, z a problem over U, and P a probability distribution for z. Let T be an arbitrary node in the graph $\Delta(z)$, and p the number assigned to T by the optimization procedure. Then for each decision tree Γ from the set $Tree_{\Psi,P}(T)$, the equality $p = N(T, P)h_\Psi(\Gamma, z_T, P)$ holds.*

Proof. Prove the lemma by induction on the nodes of $\Delta(z)$. For each terminal node T, only one irredundant decision tree Γ exists depicted on Fig. 4.1 and the statement of the lemma obviously holds for T. Let now T be a nonterminal node and the statement of lemma holds for all descendants of T. Consider an arbitrary decision tree $\Gamma \in Tree_{\Psi,P}(T)$. Let the root of Γ be labeled with an attribute f_i. Let $E(T, i) = \{a_1, \ldots, a_t\}$. For $j = 1, \ldots, t$, denote by Γ_j the decision tree connected to the root of Γ with the edge labeled with a_j. Let for $j = 1, \ldots, t$, the node $T(f_i, a_j)$ be labeled with a number p_j. For $j = 1, \ldots, t$, denote $z_j = z_{T(f_i, a_j)}$.

The induction base implies that the equality $p_j = N(T(f_i, a_j), P)h_\Psi(\Gamma_j, z_j, P)$ holds for $j = 1, \ldots, t$. According to the definition of the optimization procedure, $p = \Psi(f_i)N(T, P) + \sum_{j=1}^{t} p_j$. Since Γ is an irredundant decision tree, for any $\delta \notin E(T, i)$, the edge that leaves the root of Γ and is labeled with δ, enters a terminal node.

From the definition of the average weighted depth we have

$$h_\Psi(\Gamma, z_T, P) = \Psi(f_i) + \frac{1}{N(T, P)} \sum_{j=1}^{t} N(T(f_i, a_j))h_\Psi(\Gamma_t, z_j, P) \,. \qquad (4.1)$$

From the last three equalities we have $p = N(T, P)h_\Psi(\Gamma, z_T, P)$. Since Γ is an arbitrary tree from $Tree_{\Psi,P}(T)$, all the trees in $Tree_{\Psi,P}(T)$ have the same complexity equal to p. $\qquad\square$

Proof of Theorem 4.1. The theorem will be proved by induction on the nodes of the graph $\Delta(z)$. Let T be a terminal node. Then the set $Tree(T)$ contains only the tree depicted on Fig. 4.1 and this tree is not removed by the optimization procedure. Then the statement of the theorem holds for the node T.

Let now T be a nonterminal node in $\Delta(z)$, and the statement of the theorem hold for any descendant of T in the graph $\Delta(z)$. Let the optimization procedure assigned a number p to the node T. Lemma 4.1 implies that all decision trees in $Tree_{\Psi,P}(T)$ have the same complexity p. Consider an arbitrary decision tree Γ from the set $Tree^\star_{\Psi,P}(T)$. From the definition of the set $Tree^\star_{\Psi,P}(T)$ we have

$$N(T, P)h_\Psi(\Gamma, z_T, P) \leq p . \tag{4.2}$$

Let us show that $N(T, P)h_\Psi(\Gamma, z_T, P) = p$. Let the root of Γ be assigned with an attribute f_i. Since Γ is an irredundant decision tree, $i \in E(T)$. Let $E(T, i) = \{a_1, \dots, a_t\}$. For $j = 1, \dots, t$, denote by Γ_j the subtree that is connected to the root with the edge labeled with a_j. One can see that Γ_j is contained in the set $Tree(T(f_i, a_j))$. Let p_j be the number assigned to the node $T(f_i, a_j)$ during optimization. For $j = 1, \dots, t$, denote $z_j = z_{T(f_i, a_j)}$. Since the theorem holds for the node $T(f_i, a_j)$, we have $N(T(f_i, a_j), P)h_\Psi(\Gamma_j, z_j, P) \geq p_j$. From the description of the optimization process it follows that $\Psi(f_i)N(T, P) + \sum_{j=1}^t p_j \geq p$. Since Γ is an irredundant decision tree, for any $\delta \notin E(T, i)$, the edge that leaves the root of Γ and is labeled with δ, enters a terminal node.

From the two last equalities and (4.1) we have $N(T, P)h_\Psi(\Gamma, z_T, P) \geq p$, and, recalling (4.2), $N(T, P)h_\Psi(\Gamma, z_T, P) = p$. Then

$$Tree_{\Psi,P}(T) \subseteq Tree^\star_{\Psi,P}(T) . \tag{4.3}$$

Due to (4.1), optimality of the tree Γ implies optimality of each tree Γ_j, so $\Gamma_j \in Tree^\star_{\Psi,P}(T(f_i, a_j))$ for $j = 1, \dots, t$. Then, according to the induction base, Γ_j belongs to the set $Tree_{\Psi,P}(T(f_i, a_j))$ for $j = 1, \dots, t$. Consider the bundle of edges in the graph $\Delta(z)$ that leave the node T and are labeled with the pairs $(f_i, a_1), \dots, (f_i, a_t)$. Since $N(T, P)h_\Psi(\Gamma, z_T, P) = p$, these edges were not removed by the optimization procedure. Then, according to the definition of the set $Tree_{\Psi,P}(T)$, the tree Γ belongs to this set. As Γ was chosen arbitrarily, we have $Tree^\star_{\Psi,P}(T) \subseteq Tree_{\Psi,P}(T)$, and due to (4.3), $Tree^\star_{\Psi,P}(T) = Tree_{\Psi,P}(T)$. $\qquad\square$

Consider an algorithm $\mathcal{A}$ that given a table T_z^* first builds a graph $\Delta(z)$, then transforms it resulting graph $\Delta_{\Psi,P}(z)$ and finally extracts one of trees described by the graph $\Delta_{\Psi,P}(z)$. For an arbitrary polynomial Q, a probability distribution P is called Q-restricted if for an arbitrary row $\bar{d} \in T_z$, the length of the binary notation of the number $P(\bar{d})$ does not exceed $Q(n)$ where n is the number of columns in the table. The following theorem characterizes the time complexity of the algorithm $\mathcal{A}$.

Theorem 4.2. *Let $Q(x)$ be some polynomial. Then for an arbitrary problem $z = (\nu, f_1, \ldots, f_n)$ and an arbitrary Q-restricted probability distribution P for the problem z, the working time of the algorithm $\mathcal{A}$ is proportional to the number of rows $D(T_z)$ if the table T_z^* is terminal. If the table T_z^* is nonterminal, the working time of the algorithm $\mathcal{A}$ is bounded from below by the maximum of the values n, the number of nonterminal separable subtables $|S(z)|$, $D(T_z)$ and the maximum length of attribute weight in binary notation, and is bounded from above by a polynomial on these values.*

Proof. If the table T_z^* is terminal, the working time of the algorithm $\mathcal{B}$ is proportional to $D(T_z)$ according to Proposition 4.3. Since the graph $\Delta(z)$ contains a single node, the remaining steps of the algorithm $\mathcal{A}$ are completed in a constant time, so the statement of the theorem holds.

Let T_z^* be a nonterminal table. While calculating the number to assign to the node T_z in the graph $\Delta(z)$, the optimization procedure necessarily reads the weights of all attributes. This fact and Proposition 4.3 prove the lower bound on the working time of the algorithm $\mathcal{A}$.

Let us prove the upper bound. From Proposition 4.3 it follows that the working time of the algorithm $\mathcal{B}$ is limited from above by a polynomial on $D(T_z)$, n and $|S(z)|$. The optimization procedure performs exactly $(|S(z)|+1)$ steps. The time of computing p_i is limited from above by a polynomial on the maximum length of the attribute weight notation (denote it by l), $Q(n)$ and $D(T)$. The time of computing p given p_i is proportional to the number of bundles that is at most n. Given the graph $\Delta_{\Psi,P}(z)$, an optimal tree can be obtained by the time proportional to the number of nodes in the graph, which is limited from above by a polynomial on n and $|S(z)|$. Then the theorem statement is a consequence of the facts that both the number of steps and the complexity of each step are bounded from above by a polynomial on n, $|S(z)|$, $D(T_z)$ and l. □

4.2   Greedy Algorithms

A greedy algorithm builds a decision tree in a top-down fashion, minimizing some impurity criteria at each step. There are several impurity criteria based on information-theoretical [71, 78], statistical [8] and combinatorial [43] approaches. In this section, several impurity criteria are defined followed by a general description of greedy algorithm and experimental results that compare the average depth of decision trees built by different algorithms.

Let $U = (A, F)$ be an information system, Ψ a weight function for U, $z = (\nu, f_1, \ldots, f_n)$ a problem over an U, and P a probability distribution for z.

Let T be a separable subtable of T_z. Let the function $\nu(x)$ take l different values $\nu_1, \ldots, \nu_l$ on the rows of T. For $i = 1, \ldots, l$, denote $N_i = \sum_{\bar{d} \in T, \nu(\bar{d}) = \nu_i} P(\bar{d})$. We consider four *uncertainty measures*:

- entropy: $ent(T) = -\sum_{i=1}^{l} (N_i/N(T, P) \log_2(N_i/N(T, P)))$ (we assume $0 \log_2 0 = 0$);
- Gini index: $gini(T) = 1 - \sum_{i=1}^{l} (N_i/N(T, P))^2$;
- misclassification error: $me(T) = 1 - max_{i=1,\ldots,l} N_i/N(T, P)$;
- weighted number of unordered pairs of rows labeled with different decisions: $rt(T) = \left(N(T, P)^2 - \sum_{i=1}^{l} N_i^2 \right)/2$ (note that $rt(T) = N(T, P)^2 \times gini(T)/2$);

Let $i \in E(T)$ and $E(T, i) = \{a_1, \ldots, a_t\}$. The attribute f_i divides the table T into the subtables $T_1 = T(f_i, a_1), \ldots, T_t = T(f_i, a_t)$. We now define an *impurity function* I which assigns *impurity* $I(T, f_i)$ to this partition. Let us fix an uncertainty measure U from the set $\{ent, gini, me, rt\}$ and the type of impurity function: *sum* or *weighted-sum*. Then for the type *sum*, $I(T, f_i) = \sum_{j=1}^{t} U(T_j)$, and for the type *weighted-sum*, $I(T, f_i) = \sum_{j=1}^{t} U(T_j) N(T_j)/N(T)$. As a result, we have eight different impurity functions.

Consider an algorithm G that given representation of a problem z and a probability distribution P in the form of decision table T_z^* builds a decision tree $G(z, P)$.

Step 1. Assume $T = T_z$. Build a decision tree that contains a single node v. Let T be a terminal table. Then assign the number $\nu(\bar{\delta})$ to the node v where $\bar{\delta}$ is an arbitrary row from T. Denote $G(z, P)$ the resulted decision tree. The process G is completed.

Let T be a nonterminal table. Assign the word λ to the node v and proceed to the next step.

Let $t \geq 1$ steps have been already done. Denote Γ the tree built at the step t.

Step $(t+1)$. If none of the nodes in Γ is assigned with a word from Ω_z^*, then denote $G(z, P)$ the tree Γ. The process Γ is completed. Otherwise, choose in Γ a node w which is assigned with a word α from Ω_z^*.

Let $T\alpha$ be a terminal subtable. If $T\alpha = \emptyset$, then instead of α assign to w the number 0. If $T\alpha \neq \emptyset$, then instead of α assign to w the number $\nu(\bar{\delta})$ where $\bar{\delta}$ is an arbitrary row from $T\alpha$. Proceed to the step $(t+2)$.

Let $T\alpha$ be a nonterminal subtable. Then for each $f_i \in E(T\alpha)$, compute the value $I(T\alpha, f_i)$ and assign to the node w the attribute f_{i_0} where i_0 is the minimum $i \in \{1, \ldots, n\}$ for which $I(T, f_i)$ has the minimum value. For each $\delta \in E(T\alpha, f_{i_0})$, add to the tree Γ the node $w(\delta)$, mark this node with the word $\alpha(f_{i_0}, \delta)$, draw the edge from w to $w(\delta)$, and mark this edge with δ. Proceed to the step $(t+2)$.

Different impurity functions result in different greedy algorithms. The following experiment compares the average depth of decision trees built by these algorithms with the minimum average depth calculated by the algorithm $\mathcal{A}$.

Table 4.1 Average depth of decision trees built by different algorithms

Data set	Min. avg. depth	sum				weighted sum			
		ent	gini	rt	me	ent	gini	rt	me
adult-stretch	1.50	1.50	1.50	3.50	1.50	1.50	1.50	3.50	1.50
agaricus-lepiota	1.52	2.35	2.35	1.54	1.52	1.52	1.52	1.52	1.98
balance-scale	3.55	3.55	3.55	3.61	3.55	3.55	3.55	3.61	3.55
breast-cancer	3.24	6.36	6.36	4.06	3.30	3.49	3.70	3.30	3.35
cars	2.95	3.06	3.06	3.72	3.76	2.95	2.96	4.00	4.39
flags	2.72	9.31	9.73	3.21	2.81	3.16	3.16	2.82	2.80
hayes-roth-data	2.62	2.64	2.64	2.64	2.62	2.64	2.64	2.62	2.62
house-votes-84	3.54	5.88	6.99	5.29	3.77	3.68	3.80	3.77	3.63
lenses	1.80	1.80	1.80	3.00	3.00	3.00	1.80	3.00	3.00
lymphography	2.67	7.09	7.09	3.37	2.83	3.12	3.12	2.79	2.78
monks-1-test	2.50	4.50	4.50	2.50	2.50	2.50	2.50	2.50	2.50
monks-1-train	2.53	4.34	4.34	2.53	2.77	3.19	3.22	2.53	2.53
monks-2-test	5.30	5.33	5.33	5.37	5.54	5.40	5.40	5.54	5.54
monks-2-train	4.11	4.70	4.70	4.54	4.20	4.34	4.34	4.26	4.28
monks-3-test	1.83	4.11	2.78	2.78	1.83	1.83	2.08	1.83	1.83
monks-3-train	2.51	3.76	3.03	2.71	2.53	2.54	2.54	2.53	2.53
nursery	3.45	4.05	4.21	3.76	3.76	3.48	3.46	3.85	4.18
poker-hand-train	4.09	6.54	6.54	4.66	4.12	4.12	4.12	4.12	4.13
shuttle-landing	2.33	3.93	3.93	2.93	2.33	2.40	2.40	2.33	2.33
soybean-small	1.34	1.34	1.34	1.34	1.34	1.34	1.34	1.34	1.89
spect-test	2.95	5.93	5.55	4.93	3.48	3.04	3.34	3.47	3.44
teeth	2.78	4.39	4.52	2.78	2.83	2.83	2.78	2.83	2.83
tic-tac-toe	4.35	4.88	4.68	4.82	4.94	4.60	4.58	5.03	5.11
zoo-data	2.29	3.86	3.86	2.44	2.37	2.37	2.37	2.37	2.41
ARD		0.564	0.539	0.222	0.070	0.066	0.052	0.126	0.121

The data sets were taken from UCI Machine Learning Repository [25]. Each data set is represented as a table containing several input columns and an output (decision) column. Some data sets contain index columns that take unique value for each row. Such columns were removed. Some tables contain rows with identical values in all columns, possibly, except the decision column. In this case, each group of identical rows was replaced with a single row with common values in all input columns and the most common value in the decision column. Some tables contains missing values. Each missing value was replaced with the most common value in the corresponding column.

The resulted table was interpreted as a decision table T_z^* where input columns represent attribute values and the output column represents values of the function $\nu(x)$. We assume uniform probability distribution $P(x) \equiv 1$. As an integral performance measure we consider the *average relative deviation(ARD)*. For an approximate algorithm $\mathcal{X}$, a set of problems $\mathcal{Z} = \{z_1, \ldots, z_t\}$, and a set of probability distributions $\mathcal{P} = \{P_1, \ldots, P_t\}$,

$$ARD(\mathcal{X}, \mathcal{Z}, \mathcal{P}) = \frac{1}{t} \sum_{i=1}^{t} \frac{h_{\mathcal{X}}(z_i, P_i) - h(z_i, P_i)}{h(z_i, P_i)} \ ,$$

where $h_{\mathcal{X}}(z_i, P_i)$ is the P-average depth of the decision tree for z_i built by $\mathcal{X}$. We assume that none of the tables T_{z_i} are terminal, so $h(z_i, P_i) > 0$ for $i = 1, \ldots, t$.

Table 4.1 shows results of experiments with 24 data sets. Each row contains data set name, the minimum average depth of decision tree calculated by the algorithm $\mathcal{A}$, and the average depth of decision trees built by each of the eight greedy algorithms. The last row shows the average relative difference for the greedy algorithms. One can see that a combination of *weighted sum* with Gini index (the criterion used by CART [8]) and entropy (the criterion used by ID3 [70]) results in the least ARD values.

4.3   Modeling Monotonic Boolean Functions by Decision Trees

The property of the algorithm $\mathcal{A}$ to build optimal decision trees can be used to find exact values of the Shannon-type function $H_B(n)$ described in Chap. 3.2 for small n. In this section, an experiment is described that calculates $H_B(n)$ for monotone functions depending on up to six variables. The number of monotone functions of n arguments (also known as Dedekind number $M(n)$) is a rapidly growing sequence. The second column of Table 4.2 contains number $M(n)$ for $n = 1, \ldots, 6$. Using algorithm $\mathcal{A}$, we built optimal

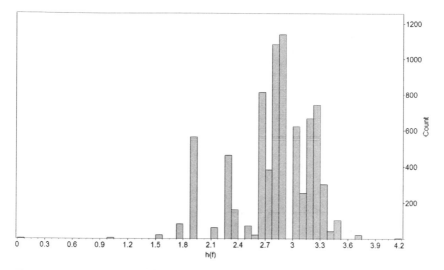

Fig. 4.3 Spectrum of $h(f)$ values for $n = 5$

decision trees for all functions and thus calculated the value of $H_B(n)$ that is given in the third column. The fourth and fifth column contain the lower and the upper bound on $H_B(n)$ which are given by Proposition 3.8.

Table 4.2 $H_B(n)$ and its bounds for the class of monotone functions

n	$M(n)$	$H_B(n)$	Lower bound	Upper bound
1	3	1	0.59	1
2	6	1.5	1.27	1.5
3	20	2.5	2	2.5
4	168	3.125	2.76	3.5
5	7561	4.125	3.55	4.5
6	7828354	4.8125	4.35	5.63

The experiments revealed that the minimum average depth reaches its maximum on threshold functions described in Sect. 3.1.3. For odd n, the only function having $h(f) = H_B(n)$ is $Thr_{n,(n+1)/2}$. For even n, there are two such functions: $Thr_{n,n/2}$ and $Thr_{n,n/2+1}$. Note that all these functions are α-functions, so the obtained value of $H_B(n)$ is the same for the classes M_1, M_2, M_3, M_4. The function $Thr_{n,(n+1)/2}$ is a self-dual function for odd n, so the obtained values of $H_B(n)$, $n = 1, 3, 5$, are applicable to the class D_2. The experiment also allows to find the histogram of distribution of the

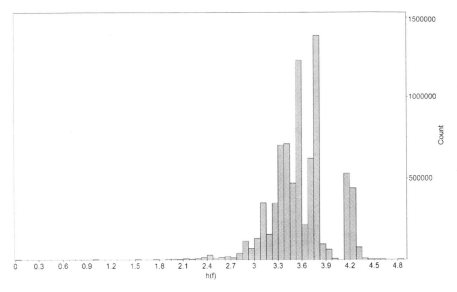

Fig. 4.4 Spectrum of $h(f)$ values for $n = 6$

minimum average depth among functions in M_1 of n variables. Fig. 4.3 and 4.4 show histogram of $h(f)$ for monotone functions of five and six variables.

4.4   Constructing Optimal Decision Trees for Corner Point Detection

In this section, we consider a problem that originated in computer vision: constructing an optimal testing strategy for corner point detection by FAST algorithm [74, 75]. The problem can be formulated as a problem of building a decision tree with the minimum average depth. We experimentally compare performance of the algorithm $\mathcal{A}$ and several greedy algorithms that differ in the attribute selection criterion.

4.4.1   Corner Point Detection Problem

One of the important problems considered in computer vision is object tracking that is given a video stream, locating an object and determining its position in each frame. There are several approaches to object tracking. One of the accepted approaches is detecting feature points and acquiring the object position by these points. Rosten and Drummond devised FAST algorithm [74, 75] that tracks an object by position of its corners and proposed a simple

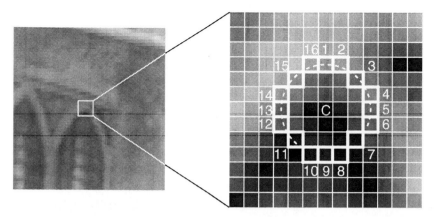

Fig. 4.5 [74] FAST Feature detection in an image patch. The highlighted squares are the pixels used in the feature detection. The pixel at C is the center of a detected corner: the dashed line passes through 12 contiguous pixels that are brighter than C by more than the threshold γ.

algorithm for corner point detection. The algorithm iterates through all image pixels and detects corner points by comparing the intensity of the current pixel and surrounding pixels. In order to determine if an image pixel a is a corner point, a circle of 16 pixels (a Bresenham circle of radius 3) surrounding a is examined: the intensity of each pixel of the circle is compared with the intensity of a. The pixel a is assumed to be a corner point if at least 12 contiguous pixels on the circle are all either brighter or darker than a by a given threshold γ (Fig. 4.5).

The surrounding pixels can be tested in an arbitrary order, and the required number of tests depends on the data and the chosen order of testing. A good testing strategy can reduce the expected number of checks and thus reduce the running time of the algorithm. One can see that in order to claim the current pixel as a corner point, at least 12 checks needs to be done, but some candidate pixels can be rejected after only four checks. For example, checking the circle points 1, 5, 9 and 13 allows to reject candidates that do not have at least three out of four pixels either darker or lighter than the central pixel.

For an arbitrary pixel a and for $i = 1, \ldots, 16$, denote by $\phi_i(a)$ the intensity of the i-th pixel in the circle surrounding a (the pixel ordering is shown on Fig. 4.3) and denote by $\phi_c(a)$ the intensity of the pixel a. The pixel a can be represented as an object that is characterized by the attributes $f_1, \ldots, f_{16}$ where

$$f_i(a) = \begin{cases} 0 \,, & \text{if } \phi_i(a) - \phi_c(a) < -\gamma \,, \\ 1 \,, & \text{if } |\phi_i(a) - \phi_c(a)| \le \gamma \,, \\ 2 \,, & \text{if } \phi_i(a) - \phi_c(a) > \gamma \,. \end{cases}$$

The problem of corner point detection can be formulated as a problem $z = (\nu, f_1, \ldots, f_{16})$ over a 3-valued information system $U = (A, F)$. The set A contains examples of image patches collected from the training data set that is a set of images or a video fragment that is similar to one the algorithm will work with. The set F consists of the attributes $f_1, \ldots, f_{16}$, and the function ν takes the value 1 if given combination of attribute values correspond to a corner point, and 0 otherwise. Using the training data set, one can also estimate a probability distribution P as cardinality of the equivalence classes obtained by partitioning of A with the attributes from F. Then a valid testing strategy can be represented by a decision tree for z that solves z, and an optimal strategy corresponds to a tree with the minimum P-average depth.

4.4.2   Experimental Results

Following the method proposed by the authors of FAST algorithm, we estimated the probability distribution from the training data. For each pixel a (except a 2-pixel outer boundary of each image), we calculated the tuple $(f_1(a), \ldots, f_{16}(a))$ of attribute values. Then we formed a decision table T_z^* that contains as rows all tuples of attribute values encountered in the training data. Each row is assigned with the estimated probability that is the number of occurrences of the corresponding tuple. We did not include to the decision table the rows that do not appear in the training data. These tuples of attribute values may encounter on other images and may be misclassified, but we suppose they are less probable, so the number of misclassifications will be small and can be compensated by the sensor fusion technique on a subsequent stage of the object tracking algorithm.

We performed an experiment that compares the average depth of decision trees built by the greedy algorithms with the minimum value. For training, we took three groups of images considered in [75] named box, junk and maze and tried five values of the threshold γ: 30, 40, 50, 70 and 100. For each set of images and each threshold value, a decision table T_z^* was constructed. Then for each decision table, decision trees were build by the algorithm $\mathcal{A}$ and by three greedy algorithms that use a combination of weighted-sum impurity function with gini, ent and me uncertainty measures respectively.

Table 4.3 Characteristics of decision tables

Data set	Threshold	# of rows	# of corner points	# of nodes	# of edges	Time, s
maze	100	1343	8	88577	363418	1
maze	70	5303	476	1699986	9223694	59
maze	50	15198	2135	5030983	24660203	339
maze	40	27295	4165	8167123	37285264	830
maze	30	50750	9310	14278124	60596561	2404
junk	100	146	0	0	0	0
junk	70	980	8	40045	157697	1
junk	50	3509	101	742765	4055185	20
junk	40	8323	282	1926830	9946555	85
junk	30	18243	847	4379006	22110004	350
box	100	680	15	58186	308882	1
box	70	3225	113	918734	4876964	23
box	50	10972	546	4059543	20901371	215
box	40	20080	1487	7075517	33320358	574
box	30	38381	4258	12404116	53458575	1660

Table 4.4 Average depth of decision trees

Data set	Threshold	Min. avg. depth γ	Uncertainty measure		
			ent	*gini*	*me*
maze	100	1.27327	1.38421	1.4073	1.28518
maze	70	2.97021	3.31982	3.43315	3.32529
maze	50	3.07119	3.25339	3.49671	3.38137
maze	40	3.13391	3.28679	3.55746	3.45605
maze	30	3.27496	3.4028	3.73163	3.53888
junk	70	1.17653	1.22653	1.22653	1.22653
junk	50	2.58393	2.73041	2.77344	2.76917
junk	40	2.77556	3.01826	2.97345	2.98738
junk	30	2.84794	3.04665	3.08738	3.17831
box	100	1.26912	2.19706	2.28824	1.28235
box	70	2.68217	2.83969	3.05891	2.8093
box	50	3.05851	3.24426	3.34962	3.35992
box	40	3.14631	3.43322	3.60652	3.60931
box	30	3.27373	3.47195	3.62302	3.72645
ARD			0.10738	0.14914	0.07744

For each decision table, Table 4.3 cites the total number of rows and the number of corner points detected in the training data. Additionally, the table shows the number of nodes and the number of edges in the graph $\Delta(z)$, and the running time of the algorithm $\mathcal{A}$. The results give an evidence that $\mathcal{A}$ is capable of processing a table containing up to $50,000$ rows and producing a graph with more than $14,000,000$ nodes. Table 4.4 gives the average depth of decision trees built by the exact algorithm and by the greedy algorithms. One can see that an optimal strategy requires less than four points to test on average. For each uncertainty measure U, the last row contains the average relative deviation of the greedy algorithm that uses U in the impurity criteria.

One can see that the greedy algorithms construct decision trees with the average depth $7 - 15\%$ greater than the minimum. The greedy algorithm that uses uncertainty measure me has the minimum ARD. However, for smaller values of γ, there is larger variety of data (more rows in the decision table) and the greedy algorithm that uses ent performs better (this is, in fact, ID3 algorithm [70] applied for this problem in [75]).

Chapter 5
Problems over Information Systems

The problems of estimation of the minimum average time complexity of decision trees and design of efficient algorithms are complex in general case. The upper bounds described in Chap. 2.4.3 can not be applied directly due to large computational complexity of the parameter $M(z)$. Under reasonable assumptions about the relation of P and NP, there are no polynomial time algorithms with good approximation ratio [12, 32]. One of the possible solutions is to consider particular classes of problems and improve the existing results using characteristics of the considered classes.

We use the notion of information system to describe a class of problems. The set of objects and the set of attributes are allowed to be infinite (but countable). Among all information systems, we distinguish the restricted information systems in which any system of equations of the type *"attribute"* = *"value"* has an equivalent subsystem whose weight is below a predefined threshold.

The first section describes the notion of restricted information system and gives bounds on the average weighted depth of decision trees depending only on the entropy. In the second section, we prove that for a restricted information system, under reasonable assumptions about weight function and probability distribution, the time complexity of the algorithm $\mathcal{A}$ is limited from above by a polynomial on the number of attributes in the problem description. Some results of this chapter were published in [17].

5.1 On Bounds on Average Depth of Decision Trees Depending Only on Entropy

Let $U = (A, F)$ be an information system and Ψ be a weight function for U. Theorem 2.3 gives a bound on the minimum average weighted depth of

I. Chikalov: Average Time Complexity of Decision Trees, ISRL 21, pp. 79–84.
springerlink.com

decision tree for an arbitrary problem z over U. However, efficiency of this
bound is limited due to large computational complexity of the parameter
$M_\Psi(z)$. Let us consider a necessary and sufficient condition for existence of a
function Φ such that $h_\Psi(z, P) \le \Phi(H(P))$ for any problem z over U and any
probability distribution P for z.

For an arbitrary natural number t, a system of equations of the form

$$\{f_1(x) = \delta_1, \ldots, f_t(x) = \delta_t\}, \tag{5.1}$$

where $f_1, \ldots, f_t \in F$ and $\delta_1, \ldots, \delta_t \in E_k$ is called *a system of equations over*
U. A system of equations over U is called *irreducible*, if it does not have any
proper equivalent subsystems. An information system U is called r-*restricted*
(restricted) if each compatible system of equations over U has an equivalent
subsystem that contains at most r equations.

For the system of equations (5.1), the value $\sum_{i=1}^{t} \Psi(f_i)$ is called *the weight
of the system*. An information system U is called r-*restricted (restricted)
relative to* Ψ if each compatible system of equations over U has an equivalent
subsystem whose weight does not exceed r.

Example 5.1. Let $A = R^n$, and F be a nonempty set of mappings from R^n to
R. Consider an infinite family of functions $[F] = \{\text{sign}(f + \alpha) + 1 : f \in F, \alpha \in$
$R\}$ (note that the expression $(\text{sign}(x) + 1)$ takes the value 0 for a negative x,
1 for $x = 0$, and 2 for a positive x). If $|F| = k < \infty$, then the information
system $U = (A, [F])$ is $2k$-restricted (or $2k$-restricted relative to the weight
function $\Psi \equiv 1$).

The following theorem for an arbitrary problem over a restricted information
system and an arbitrary probability distribution, gives an upper bound on
the minimum average weighted depth of decision tree that depends only on
the entropy of probability distribution.

Theorem 5.1. *Let U be an information system, Ψ a weight function for U
and U be r-restricted relative to Ψ where r is some natural number. Then
$h_\Psi(z, P) \le 2r(H(P) + 1)$ for an arbitrary problem z over U and an arbitrary
probability distribution P for z.*

Proof. Let U be a k-valued information system, $U = (A, F)$ and $z =$
$(\nu, f_1, \ldots, f_n)$. If $z \equiv \text{const}$ on A, then obviously $M_\Psi(z) \le r$. Let $z \not\equiv \text{const}$ on
A. Let us consider an arbitrary tuple $\bar{\delta} = (\delta_1, \ldots, \delta_n)$ from E_k^n and show that
$M_\Psi(z, \bar{\delta}) \le 2r$. From the definition of the parameter $M_\Psi(z, \bar{\delta})$ it follows that
there exists an irreducible system of equations $S = \{f_{i_1}(x) = \delta_{i_1}, \ldots, f_{i_t}(x) =$
$\delta_{i_t}\}$ over z such that $t > 0$ and the weight of the system is $M_\Psi(z, \bar{\delta})$. Denote by $A(S)$ the set of solutions of this system on A. If $A(S) \ne \emptyset$, then the

weight of S does not exceed r. Let $A(S) = \emptyset$. Denote $S_1 = S \setminus \{f_{i_1}(x) = \delta_{i_1}\}$. The equality $A(S) = \emptyset$ and the fact that S is irreducible imply that S_1 is a compatible irreducible system. Therefore, the weight of the system S_1 does not exceed r and the weight of the system S does not exceed $r + \Psi(f_{i_1})$. According to the definition of system of equations, $f_{i_1}(x) \not\equiv \text{const}$ on A. Consequently, there exists a number $\delta \in E_k$ for which the set of solutions of the equation $f_{i_1}(a) = \delta$ is a nonempty proper subset of A. Then the system of equations $\{f_{i_1}(x) = \delta\}$ is irreducible and compatible, and its weight (that is equal to $\Psi(f_{i_1})$) does not exceed r. Therefore, the weight of the system of equations S does not exceed $2r$ and the value of the parameter $M_\Psi(z)$ (as the maximum of $M_\Psi(z, \bar{\delta})$, $\bar{\delta} \in E_k^n$) does not exceed $2r$. Theorem 2.3 implies that $h_\Psi(z, P) \leq 2r(H(P) + 1)$. $\qquad\square$

The following theorem shows that the conditions of Theorem 5.1 are necessary and sufficient for existence of a linear upper bound depending only on the entropy and considering non-linear bounds does not extend the class of information systems that have upper bounds depending only on the entropy.

Theorem 5.2. *Let U be an information system that is not restricted relative to the weight function Ψ for U. Then for an arbitrary $\varepsilon > 0$, there is no function Φ that is limited within the interval $[0, \varepsilon]$ and possesses the condition $h_\Psi(z, P) \leq \Phi(H(P))$ for any problem z over U and any probability distribution P for z.*

Proof. Let $U = (A, F)$ be a k-valued information system. Assume that for some $\varepsilon > 0$, there exists a function Φ such that $\Phi(x) \leq K$, $x \in [0, \varepsilon]$, and $h_\Psi(z, P) \leq \Phi(H(P))$ for any problem z over U and any probability distribution P for z. By the premise of the theorem, for each natural number i, there exists an irreducible compatible system of equations S_i with the weight at least i. For $i = 1, 2, \ldots$, set into correspondence to the system of equations S_i a problem z_i over U. Let $S_i = \{f_1^i(x) = \delta_1^i, \ldots, f_{n_i}^i(x) = \delta_{n_i}^i\}$. Then $z_i = (\nu_i, f_1^i, \ldots, f_{n_i}^i)$ where $\nu_i : \{0, 1\}^{n_i} \to \omega$,

$$\nu_i(\bar{\delta}) = \begin{cases} 1, & \text{if } \bar{\delta} = \bar{\delta}^i, \\ 0, & \text{if } \bar{\delta} \neq \bar{\delta}^i, \end{cases}$$

and $\bar{\delta}^i = (\delta_1^i, \ldots, \delta_{n_i}^i)$. Let the table T_{z_i} contain s_i rows. Define a probability distribution P_i as follows:

$$P_i(\bar{d}) = \begin{cases} (s_i^2 - s_i + 1), & \text{if } \bar{d} = \bar{\delta}^i, \\ 1, & \text{if } \bar{d} \neq \bar{\delta}^i. \end{cases}$$

Let the function Ψ be not limited. This allows for choosing the systems S_i such that each of them consists of a single equation. Then $s_i \in \{2, \ldots, k\}$ for any $i \in \omega \setminus \{0\}$, and the value $\Phi(H(P_i))$ takes at most k different values. Consequently, $\Phi(H(P_i))$ is limited from above by some constant (for convenience of notations let it be equal to K).

Let the function Ψ be limited. This allows for choosing the systems of equations such that for any natural i the system S_i contains at least i equations. Irreducibility of the system S_i implies the inequality $s_i \geq i$. From the definition of entropy it follows that $H(P_i) \geq 0$ for any natural number i. Apply the following transformations:

$$H(P_i) = \log_2 s_i^2 - \frac{1}{s_i^2}(s_i^2 - s_i + 1)\log_2(s_i^2 - s_i + 1)$$

$$= \log_2 s_i^2 - \left(1 - \frac{s_i - 1}{s_i^2}\right)\left(\log_2 s_i^2 + \log_2(1 - \frac{s_i - 1}{s^2})\right)$$

$$< -\left(1 - \frac{s_i - 1}{s_i^2}\right)\log_2\left(1 - \frac{s_i - 1}{s_i^2}\right) + \frac{2}{s_i}\log_2 s_i \; .$$

One can see that for $i \to \infty$, both summands tend to zero. Then there exists a number $i_0 \in \omega \setminus \{0\}$ such that $H(P_i) < \varepsilon$ for $i \geq i_0$. According to the assumption, $\Phi(H(P_i)) \leq K$ for $i \geq i_0$.

Let Γ_i be a decision tree for the problem z_i that solves z_i, and Γ_i be optimal for Ψ, z_i and P_i. Then there exists a complete path ξ_i in Γ_i such that $\bar{\delta}^i \in T_{z_i}\pi(\xi_i)$. Since for an arbitrary row $\bar{d} \in T_{z_i}$, $\bar{d} \neq \bar{\delta}_i$, the relation $\nu(\bar{d}) \neq \nu(\bar{\delta}_i)$ holds, the subtable $T_{z_i}\pi(\xi_i)$ does not contain other rows except $\bar{\delta}_i$. Irreducibility of the system S_i implies $\Psi(\xi_i) \geq i$, and the nonterminal nodes of the path ξ_i are assigned with all attributes from the set $\{f_1^i, \ldots, f_{n_i}^i\}$. Using the definition of the average weighted depth, we obtain

$$h_\Psi(\Gamma, P_i) \geq \frac{\Psi(\xi_i)P_i(\delta^i)}{N(T_{z_i}, P_i)} \geq \frac{i(s_i^2 - s_i + 1)}{s_i^2} \geq \frac{i}{2} \; .$$

Taking into account that the tree Γ_i is optimal for Ψ, z_i and P_i, we have $h_\Psi(z_i, P_i) \geq i/2$. Therefore, there exists a number $i_1 \in \omega \setminus \{0\}$ such that $h_\Psi(z_i, P_i) > K$ for $i \geq i_1$. Then for any number $i^* > \max(i_0, i_1)$, the inequality $h_\Psi(z_{i^*}, P_{i^*}) > \Phi(H(P_{i^*}))$ holds. Consequently, the considered assumption is wrong. $\qquad\square$

5.2   Polynomiality Criterion for Algorithm $\mathcal{A}$

Let $U = (A, f_1, f_2, \ldots)$ be an infinite information system and Ψ a weight function for U. Denote $\mathcal{Z}(U)$ the set of problems over the information system U. For an arbitrary problem z, denote by $\dim z$ the number of attributes listed in the description of z.

Consider the functions

$$\mathcal{S}_U(n) = \max\{|S(z)| : z \in \mathcal{Z}(U), \dim z \le n\}$$

and

$$\mathcal{D}_U(n) = \max\{D(T_z) : z \in \mathcal{Z}(U), \dim z \le n\}$$

that characterize the dependence of the maximum number of separable subtables and the maximum number of rows on the number of columns in decision tables over U.

Let Ψ be restricted from above by some constant, and $Q(x)$ be some polynomial. Theorem 4.2 implies that for an arbitrary problem z over U and an arbitrary Q-restricted probability distribution for the problem z, the time complexity of the algorithm $\mathcal{A}$ is restricted from above by a polynomial on the number of attributes in the problem description if the functions $\mathcal{S}_U(n)$ and $\mathcal{D}_U(n)$ are restricted from above by a polynomial on n. Also, one can see that the time complexity of the algorithm $\mathcal{A}$ has an exponential lower bound if the function $\mathcal{S}_U(n)$ grows exponentially.

Theorem 5.3. *Let $U = (A, F)$ be a k-valued information system. Then the following statements hold:*

a) if U is r-restricted, then $\mathcal{S}_U(n) \le (nk)^r + 1$ and $\mathcal{D}_U(n) \le (nk)^r + 1$ for any natural number n;

b) if U is not restricted, then $\mathcal{S}_U(n) \ge 2^n - 1$ for any natural number n.

Proof. a) Let U be r-restricted and $z = (\nu, f_1, \ldots, f_n) \in \mathcal{Z}(U)$. One can see that the values $|S(T_z)|$ and $D(T_z)$ do not exceed the number of pairwise nonequivalent compatible subsystems of the system of equations $\{f_1(x) = 0, \ldots, f_n(x) = 0, \ldots, f_1(x) = k - 1, \ldots, f_n(x) = k - 1\}$ including the empty system (assume the set of solutions of the empty system to be equal to A). Since the system of equations U is r-restricted, each compatible system of equations over U contains an equivalent subsystem of at most r equations. Then $|S(z)| \le (\dim z)^r k^r + 1$ and $D(T_z) \le (\dim z)^r k^r + 1$. Therefore, $\mathcal{S}_U(n) \le (nk)^r + 1$ and $\mathcal{D}_U(n) \le (nk)^r + 1$.

b) Let U be not restricted and n be a natural number. Then there exists an irreducible system of equations over U containing at least n. equations. Since each its subsystem is irreducible, there exists an irreducible system over U that consists of n equations. Let it be the system (5.1) which will be denoted by W. Let us show that any two different subsystems W_1 and W_2 of W are not equivalent. Assume the contrary. Then the systems $W \setminus (W_1 \setminus W_2)$ and $W \setminus (W_2 \setminus W_1)$ are equivalent to W, and at least one of them is a proper subsystem of W that is impossible. Consider a diagnostic problem $z = (\nu, f_1, \ldots, f_n)$. One can set into correspondence to a proper subsystem $\{f_{i_1}(x) = \delta_1, \ldots, f_{i_t}(x) = \delta_t\}$ of the system W the separable subtable $T_z(f_{i_1}, \delta_1) \ldots (f_{i_t}, \delta_t)$ of the table T_z. Since any two different subsystems are nonequivalent to each other and to the system W, the subtables corresponding to these subsystems are different and nonterminal. Then $|S(T_z)| \geq 2^n - 1$. Therefore, $\mathcal{S}_U(n) \geq 2^n - 1$. $\qquad\square$

Conclusions

The monograph considers several aspects of the problem of constructing decision trees with the minimum time complexity. A known bound on the minimum average depth for a problem with a complete set of attributes is generalized in two ways. First, a bound for an arbitrary problem is obtained that depends on the parameter $M(z)$. Second, a class of restricted information systems is described; so all problems over a restricted information system have a common bound depending only on the entropy. A necessary condition for the problem decomposition is described that might be too restrictive for using in applications, but it works in constructive proofs.

An exact algorithm $\mathcal{A}$ for construction of decision trees has been studied both theoretically and experimentally. The experimental results described in Chap. 4.4.2 show that $\mathcal{A}$ is capable of processing a table with 16 attributes and more than 50000 rows. A class of all information systems was described for which the algorithm has polynomial time complexity on the decision table size. It allows further optimization by using branch and bound methods to reduce the search space. Current parallel computing environments enable an effective implementation of such algorithms and make it applicable for practical problems described by decision tables of a moderate size.

Appendix A
Closed Classes of Boolean Functions

The lattice of all classes of Boolean functions closed relative to the operation of substitution has been described by Post in [67, 68]. In [36], Yablonskii, Gavrilov and Kudriavtzev considered the structure of all classes of Boolean functions closed relative to the operation of substitution and the operations of insertion and deletion of inessential variable. Appendix contains the description of this structure that is slightly different from Post's lattice. The text of Appendix is close to the text of Appendix in [50].

A.1 Some Definitions and Notation

Let U be a set of Boolean functions, $f(x_1, \ldots, x_n)$ be a function from U, and g_i be either a function from U or a variable, $i = 1, \ldots, n$. We will say that the function $f(g_1, \ldots, g_n)$ is obtained from functions from U *by the operation of substitution*.

Let $f(x_1, \ldots, x_n)$ be a Boolean function. A variable x_i of the function f is *essential* if there exist two n-tuples $\bar{\delta}$ and $\bar{\sigma}$ from E_2^n that differ only in the i-th digit and for which $f(\bar{\delta}) \neq f(\bar{\sigma})$. The variables of the function f that are not essential are called *inessential* variables. Let x_j be an inessential variable of the function f and $g(x_1, \ldots, x_{j-1}, x_{j+1}, \ldots, x_n) = f(x_1, \ldots, x_{j-1}, 0, x_{j+1}, \ldots, x_n)$. We will say that the function g is obtained from f *by the operation of deletion of inessential variable*. We will say that the function f is obtained from g *by the operation of insertion of inessential variable*.

Let U be a nonempty set of Boolean functions. We denote by $[U]$ the closure of the set U relative to the operation of substitution and the operations of insertion and deletion of inessential variable. The set U is called *a closed class* if $U = [U]$.

I. Chikalov: Average Time Complexity of Decision Trees, ISRL 21, pp. 87–85.
springerlink.com   © Springer-Verlag Berlin Heidelberg 2011

The notion of *a formula over* U is defined inductively in the following way:

a) The expression $f(x_1, \ldots, x_n)$, where $f(x_1, \ldots, x_n)$ is a function from U, is a formula over U.

b) Let $f(x_1, \ldots, x_n)$ be a function from U and $\varphi_1, \ldots, \varphi_n$ be expressions that are either formulas over U or variables. Then the expression $f(\varphi_1, \ldots, \varphi_n)$ is a formula over U.

A Boolean function corresponds in natural way to any formula over U. We will say that the formula *realizes* this Boolean function.

Denote by $[U]_1$ the closure of the set U relative to the operation of substitution. One can show that $[U]_1$ coincides with the set of functions realized by formulas over U. Denote by $[U]_2$ the closure of the set $[U]_1$ relative to the operations of insertion and deletion of inessential variable. One can show that $[U] = [U]_2$.

We denote the logical negation operation by $\neg$ and the modulo 2 summation by $\oplus$. For a natural n and $t \in E_2$, denote by $\tilde{t}_n$ the n-tuple $(t, t, \ldots, t) \in E_2^n$. Let f be a Boolean function depending on n variables. The function f is called α-*function* if $f(\tilde{t}_n) = t$ for any $t \in E_2$, β-*function* if $f(\tilde{t}_n) = 1$ for any $t \in E_2$, and γ-*function* if $f(\tilde{t}_n) = 0$ for any $t \in E_2$.

A function f is called *a linear* function if $f = c_0 \oplus c_1 x_1 \oplus \ldots \oplus c_n x_n$ where $c_i \in E_2$, $0 \le i \le n$. A function f is called *a self-dual* function if $f(x_1, \ldots, x_n) = \neg f(\neg x_1, \ldots, \neg x_n)$. A function f is called *a monotone* function if for any n-tuples $\bar{\delta} = (\delta_1, \ldots, \delta_n)$ and $\bar{\sigma} = (\sigma_1, \ldots, \sigma_n)$ from E_2^n such that $\delta_i \le \sigma_i$, $1 \le i \le n$, the inequality $f(\bar{\delta}) \le f(\bar{\sigma})$ holds.

Let $\mu \in \omega \setminus \{0, 1\}$. We will say that a function $f(x_1, \ldots, x_n)$ satisfies *the condition* $\langle a^\mu \rangle$ if for any μ tuples from E_2^n on which f takes the value 0 there exists a number $j \in \{1, \ldots, n\}$ such that in each of the considered tuples the j-th digit is equal to 0. We will say that the function f satisfies *the condition* $\langle a^\infty \rangle$ if there exists a number $j \in \{1, \ldots, n\}$ such that in any n-tuple from E_2^n on which f takes the value 0 the j-th digit is equal to 0. We will say that the function f satisfies *the condition* $\langle A^\mu \rangle$ if for any μ tuples from E_2^n on which f takes the value 1 there exists a number $j \in \{1, \ldots, n\}$ such that in each of the considered tuples the j-th digit is equal to 1. We will say that a function f satisfies *the condition* $\langle A^\infty \rangle$ if there exists a number $j \in \{1, \ldots, n\}$ such that in any n-tuple from E_2^n on which f takes the value 1 the j-th digit is equal to 1. The constant 1, by definition, satisfies the condition $\langle a^\infty \rangle$ and does not satisfy the condition $\langle A^2 \rangle$. The constant 0, by definition, satisfies the condition $\langle A^\infty \rangle$ and does not satisfy the condition $\langle a^2 \rangle$.

Let $\mu \in \omega \setminus \{0, 1\}$. Denote

$$h_\mu = \bigvee_{i=1}^{\mu+1} (x_1 \wedge x_2 \wedge \ldots \wedge x_{i-1} \wedge x_{i+1} \wedge \ldots \wedge x_{\mu+1})$$

and

$$h_\mu^* = \bigwedge_{i=1}^{\mu+1} (x_1 \vee x_2 \vee \ldots \vee x_{i-1} \vee x_{i+1} \vee \ldots \vee x_{\mu+1}) \,.$$

A.2   Description of All Closed Classes of Boolean Functions

In this subsection, all closed classes of Boolean functions are listed. For each class the Post notation is given, the description of functions contained in the considered class is presented, and a finite set of Boolean functions is given such that its closure relative to the operation of substitution and the operations of insertion and deletion of inessential variable is equal to this class.

As in [36], two Boolean functions are called *equal* if one of them can be obtained from the other by the operations of insertion and deletion of inessential variable.

The inclusion diagram for closed classes of Boolean functions [36] is depicted in Fig. A.1. Each closed class is represented by a dot. The dots corresponding to certain classes U and V are connected with an edge if V is immediately included into U (there are no intermediate classes between U and V); in this case, the dot corresponding to the outer class U is placed higher on the diagram. 1. The class $O_1 = [\{x\}]$. This class consists of all functions equal to the function x, and all functions obtained from them by renaming of variables without identification.

2. The class $O_2 = [\{1\}]$. This class consists of all functions equal to the function 1.

3. The class $O_3 = [\{0\}]$. This class consists of all functions equal to the function 0.

4. The class $O_4 = [\{\neg x\}]$. This class consists of all functions equal to the functions x or $\neg x$, and all functions obtained from them by renaming of variables without identification.

5. The class $O_5 = [\{x, 1\}]$. This class consists of all functions equal to the functions 1 or x, and all functions obtained from them by renaming of variables without identification.

6. The class $O_6 = [\{x, 0\}]$. This class consists of all functions equal to the functions 0 or x, and all functions obtained from them by renaming of variables without identification.

7. The class $O_7 = [\{0, 1\}]$. This class consists of all functions equal to the functions 0 or 1.

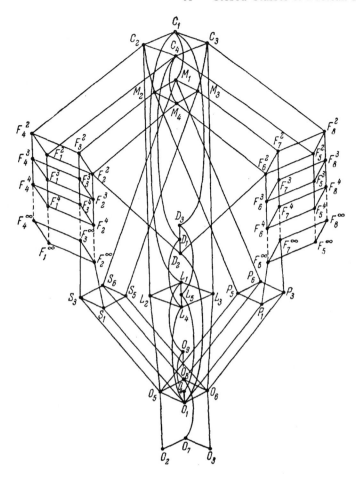

Fig. A.1 Inclusion diagram for closed classes of Boolean functions

8. The class $O_8 = [\{x, 0, 1\}]$. This class consists of all functions equal to the functions 0, 1 or x, and all functions obtained from them by renaming of variables without identification.

9. The class $O_9 = [\{\neg x, 0\}]$. This class consists of all functions equal to the functions 0, 1, $\neg x$, or x, and all functions obtained from them by renaming of variables without identification.

10. The class $S_1 = [\{x \vee y\}]$. This class consists of all disjunctions (i.e., functions of the form $\bigvee_{i=1}^{n} x_i$, $n = 1, 2, \ldots$ and all functions obtained from them by renaming of variables without identification).

11. The class $S_3 = [\{x \vee y, 1\}]$. This class consists of all disjunctions and all functions equal to 1.

12. The class $S_5 = [\{x \vee y, 0\}]$. This class consists of all disjunctions and all functions equal to 0.

13. The class $S_6 = [\{x \vee y, 0, 1\}]$. This class consists of all disjunctions and all functions equal to the functions 0 or 1.

14. The class $P_1 = [\{x \wedge y\}]$. This class consists of all conjunctions (i.e., functions of the form $\bigwedge_{i=1}^{n} x_i$, $n = 1, 2, \ldots$ and all functions obtained from them by renaming of variables without identification).

15. The class $P_3 = [\{x \wedge y, 0\}]$. This class consists of all conjunctions and all functions equal to 0.

16. The class $P_5 = [\{x \wedge y, 1\}]$. This class consists of all conjunctions and all functions equal to 1.

17. The class $P_6 = [\{x \wedge y, 0, 1\}]$. This class consists of all conjunctions and all functions equal to 0 or 1.

18. The class $L_1 = [\{x \oplus y, 1\}]$. This class consists of all linear functions.

19. The class $L_2 = [\{x \oplus y \oplus 1\}]$. This class consists of all linear α-functions and β-functions (i.e., functions of the form $\bigoplus_{i=1}^{2k} x_i \oplus 1$, $\bigoplus_{i=1}^{2l+1} x_i$, $k, l = 0, 1, 2, \ldots$ and all functions obtained from them by renaming of variables without identification).

20. The class $L_3 = [\{x \oplus y\}]$. This class consists of all linear α-functions and γ-functions (i.e., functions of the form $\bigoplus_{i=1}^{l} x_i, l = 0, 1, 2, \ldots$ and all functions obtained from them by renaming of variables without identification).

21. The class $L_4 = [\{x \oplus y \oplus z\}]$. This class consists of all linear α-functions (i.e., functions of the form $\bigoplus_{i=1}^{2l+1} x_i$, $l = 0, 1, 2, \ldots$ and all functions obtained from them by renaming of variables without identification).

22. The class $L_5 = [\{x \oplus y \oplus z \oplus 1\}]$. This class consists of all linear self-dual functions (i.e., functions of the form $\bigoplus_{i=1}^{2l+1} x_i \oplus 1, \bigoplus_{i=1}^{2l+1} x_i$, $l = 0, 1, 2, \ldots$ and all functions obtained from them by renaming of variables without identification).

23. The class $D_2 = [\{(x \wedge y) \vee (x \wedge z) \vee (y \wedge z)\}]$. This class consists of all self-dual monotone functions.

24. The class $D_1 = [\{(x \wedge y) \vee (x \wedge \neg z) \vee (y \wedge \neg z)\}]$. This class consists of all self-dual α-functions.

25. The class $D_3 = [\{(x \wedge \neg y) \vee (x \wedge \neg z) \vee (\neg y \wedge \neg z)\}]$. This class consists of all self-dual functions.

26. The class $A_1 = M_1 = [\{x \wedge y, x \vee y, 0, 1\}]$. This class consists of all monotone functions.

27. The class $A_2 = M_2 = [\{x \wedge y, x \vee y, 1\}]$. This class consists of all monotone α-functions and β-functions.

28. The class $A_3 = M_3 = [\{x \wedge y, x \vee y, 0\}]$. This class consists of all monotone α-functions and γ-functions.

29. The class $A_4 = M_4 = [\{x \wedge y, x \vee y\}]$. This class consists of all monotone α-functions.

30. The class $C_1 = [\{\neg(x \wedge y)\}]$. This class consists of all Boolean functions.

31. The class $C_2 = [\{x \vee y, x \oplus y \oplus 1\}]$. This class consists of all α-functions and β-functions.

32. The class $C_3 = [\{x \wedge y, x \oplus y\}]$. This class consists of all α-functions and γ-functions.

33. The class $C_4 = [\{x \vee y, x \wedge (y \oplus z \oplus 1)\}]$. This class consists of all α-functions.

34. The class $F_1^\mu = [\{x \vee (y \wedge \neg z), h_\mu^*\}]$, $\mu = 2, 3, \dots$. This class consists of all α-functions satisfying the condition $\langle a^\mu \rangle$.

35. The class F_2^μ, $\mu = 2, 3, \dots$ where $F_2^\mu = [\{x \vee (y \wedge z), h_2^*\}]$ if $\mu = 2$, and $F_2^\mu = [\{h_\mu^*\}]$ if $\mu \geq 3$. This class consists of all monotone α-functions satisfying the condition $\langle a^\mu \rangle$.

36. The class $F_3^\mu = [\{1, h_\mu^*\}]$, $\mu = 2, 3, \dots$. This class consists of all monotone functions satisfying the condition $\langle a^\mu \rangle$.

37. The class $F_4^\mu = [\{x \vee \neg y, h_\mu^*\}]$, $\mu = 2, 3, \dots$. This class consists of all functions satisfying the condition $\langle a^\mu \rangle$.

38. The class $F_5^\mu = [\{x \wedge (y \vee \neg z), h_\mu\}]$, $\mu = 2, 3, \dots$. This class consists of all α-functions satisfying the condition $\langle A^\mu \rangle$.

39. The class F_6^μ, $\mu = 2, 3, \dots$ where $F_6^\mu = [\{x \wedge (y \vee z), h_2\}]$ if $\mu = 2$, and $F_6^\mu = [\{h_\mu\}]$ if $\mu \geq 3$. This class consists of all monotone α-functions satisfying the condition $\langle A^\mu \rangle$.

40. The class $F_7^\mu = [\{0, h_\mu\}]$, $\mu = 2, 3, \dots$. This class consists of all monotone functions satisfying the condition $\langle A^\mu \rangle$.

41. The class $F_8^\mu = [\{x \wedge \neg y, h_\mu\}]$, $\mu = 2, 3, \dots$. This class consists of all functions satisfying the condition $\langle A^\mu \rangle$.

42. The class $F_1^\infty = [\{x \vee (y \wedge \neg z)\}]$. This class consists of all α-functions satisfying the condition $\langle a^\infty \rangle$.

43. The class $F_2^\infty = [\{x \vee (y \wedge z)\}]$. This class consists of all monotone α-functions satisfying the condition $\langle a^\infty \rangle$.

44. The class $F_3^\infty = [\{1, x \vee (y \wedge z)\}]$. This class consists of all monotone functions satisfying the condition $\langle a^\infty \rangle$.

45. The class $F_4^\infty = [\{x \vee \neg y\}]$. This class consists of all functions satisfying the condition $\langle a^\infty \rangle$.

46. The class $F_5^\infty = [\{x \wedge (y \vee \neg z)\}]$. This class consists of all α-functions satisfying the condition $\langle A^\infty \rangle$.

47. The class $F_6^\infty = [\{x \wedge (y \vee z)\}]$. This class consists of all monotone α-functions satisfying the condition $\langle A^\infty \rangle$.

48. The class $F_7^\infty = [\{0, x \wedge (y \vee z)\}]$. This class consists of all monotone functions satisfying the condition $\langle A^\infty \rangle$.

49. The class $F_8^\infty = [\{x \wedge \neg y\}]$. This class consists of all functions satisfying the condition $\langle A^\infty \rangle$.

References

1. Ahlswede, R., Wegener, I.: Suchprobleme. Teubner Verlag, Stuttgart (1979) (in German)
2. Akers, S.B.: Binary decision diagrams. IEEE Trans. Comput. 27, 509–516 (1978)
3. Alekhnovich, M., Braverman, M., Feldman, V., Klivans, A.R., Pitassi, T.: Learnability and automatizability. In: Proceedings of the 45th Annual IEEE Symposium on Foundations of Computer Science, pp. 621–630. IEEE Computer Society, Washington, DC, USA (2004)
4. Angluin, D.: Queries revisited. Theor. Comput. Sci. 313, 175–194 (2004)
5. Brace, K.S., Rudell, R.L., Bryant, R.E.: Efficient implementation of a BDD package. In: Proceedings of the 27th ACM/IEEE Design Automation Conference, DAC 1990, pp. 40–45. ACM, New York (1990)
6. Brandman, Y., Orlitsky, A., Hennessy, J.: A spectral lower bound technique for the size of decision trees and two-level AND/OR circuits. IEEE Trans. Comput. 39, 282–287 (1990)
7. Breiman, L.: Random forests. Mach. Learn. 45, 5–32 (2001)
8. Breiman, L., et al.: Classification and Regression Trees. Chapman & Hall, New York (1984)
9. Breitbart, Y., Reiter, A.: A branch-and-bound algorithm to obtain an optimal evaluation tree for monotonic Boolean functions. Acta Inf. 4, 311–319 (1975)
10. Bryant, R.: Graph-based algorithms for Boolean function manipulation. IEEE Trans. Comp. C-35(8), 677–691 (1986)
11. Cerny, E., Mange, D., Sanchez, E.: Synthesis of minimal binary decision trees. IEEE Trans. Comput. 28, 472–482 (1979)
12. Chakaravarthy, V.T., Pandit, V., Roy, S., Awasthi, P., Mohania, M.: Decision trees for entity identification: approximation algorithms and hardness results. In: Proceedings of the 26-th ACM SIGMOD-SIGACT-SIGART Symposium on Principles of Database Systems, PODS 2007, pp. 53–62. ACM, New York (2007)
13. Chang, H.Y.: An algorithm for selecting on optimum self-diagnostic tests. IEEE Transactions on Electronic Computers EC-14(5), 706–711 (1965)
14. Chegis, I., Jablonsky, S.: Logical methods for controlling electric circuit function. Proceedings of V.A. Steklov Inst. of Maths 51, 270–360 (1958) (in Russian)
15. Cheng, L., Chen, D., Wong, M.D.F.: DDBDD: delay-driven BDD synthesis for FPGAs. In: Proceedings of the 44th Annual Design Automation Conference, DAC 2007, pp. 910–915. ACM, New York (2007)

16. Chikalov, I.: Lower and upper bounds on minimal average depth of decision trees. Tech. Rep. 96-11(232), Institute of Informatics of Warsaw University, Warsaw, Poland (1996); Abstracts of the minisemester "Logic, Algebra and Computer Science, Helena Rasiova in memoriam"

17. Chikalov, I.: Bounds on average weighted depth of decision trees depending only on entropy. In: Proceedings of the Seventh International Conference on Information Processing and Management of Uncertainty in Knowledge-Based Systems, pp. 1190–1194. La Sorbonne, Paris (1998)

18. Chikalov, I.: On decision trees with minimal average depth. In: Polkowski, L., Skowron, A. (eds.) RSCTC 1998. LNCS (LNAI), vol. 1424, pp. 506–512. Springer, Heidelberg (1998)

19. Chikalov, I.: On average time complexity of decision trees and branching programs. Fundam. Inf. 39, 337–357 (1999)

20. Chikalov, I.: Algorithm for constructing of decision trees with minimal average depth. In: Proceedings of the 8th Conference on Information Processing and Management of Uncertainty in Knowledge-Based Systems, Madrid, Spain, vol. 1, pp. 376–379 (2000)

21. Chikalov, I., Moshkov, M., Zelentsova, M.S.: On optimization of decision trees. In: Peters, J., Skowron, A. (eds.) Transactions on Rough Sets IV. LNCS, vol. 3700, pp. 18–36. Springer, Heidelberg (2005)

22. Cox Jr., L.A., Qiu, Y., Kuehner, W.: Heuristic least-cost computation of discrete classification functions with uncertain argument values. Ann. Oper. Res. 21, 1–30 (1990)

23. Dorfman, R.: The detection of defective members of large populations. The Annals of Mathematical Statistics 14(4), 436–440 (1943)

24. Egler, J.F.: A procedure for converting logic table conditions into an efficient sequence of test instructions. Commun. ACM 6, 510–514 (1963)

25. Frank, A., Asuncion, A.: UCI machine learning repository (2010), http://archive.ics.uci.edu/ml

26. Friedman, J.H.: Greedy function approximation: A gradient boosting machine. Annals of Statistics 29, 1189–1232 (2000)

27. Garey, M.R.: Optimal binary identification procedures. SIAM Journal on Applied Mathematics 23(2), 173–186 (1972)

28. Garey, M.R., Graham, R.L.: Performance bounds on the splitting algorithm for binary testing. Acta Inf. 3, 347–355 (1974)

29. Gavrilov, G.P., Sapozhenko, A.A.: Problems and Exercises in Discrete Mathematics. Kluwer Texts in the Mathematical Sciences, vol. 14. Kluwer Academic Publishers, Dordrecht (2010)

30. Gower, J.C., Payne, R.W.: A comparison of different criteria for selecting binary tests in diagnostic keys. Biometrika 62(3), 665–672 (1975)

31. Gyllenberg, H.: A general method for deriving determination schemes for random collections of microbial isolates. Ann. Acad. Sci. Fenn (A)69, 1–23 (1963)

32. Heeringa, B., Adler, M.: Approximating optimal binary decision trees. Tech. Rep. 05-25, University of Massachusetts, Amherst (2005)

33. Hegedűs, T.: Generalized teaching dimensions and the query complexity of learning. In: Proceedings of the Eighth Annual Conference on Computational Learning Theory, COLT 1995, pp. 108–117. ACM, New York (1995)

34. Humby, E.: Programs from Decision Tables. Macdonald and Co., American Elsevier, London, New York (1973)

35. Hyafil, L., Rivest, R.: Constructing optimal binary decision trees is NP-complete. Information Processing Letters 5, 15–17 (1976)

36. Jablonsky, S.W., Gawrilow, G.P., Kudrjawzev, V.B.: Boolesche Funktionen und Postsche Klassen. Akademie-Verlag, Berlin (1970) (in German)

37. Kletsky, E.J.: An application of the information theory approach to failure diagnosis. IRE Transactions on Reliability and Quality Control, RQC-9(3), 29–39 (1960)

38. Knuth, D.E.: Sorting and Searching, The art of computer programming, 2nd edn. vol. 3. Addison-Wesley, Reading (1998)

39. Kospanov, E.S.: An algorithm for the construction of sufficiently simple tests. Discrete Analysis 8, 43–47 (1966) (in Russian)

40. Lee, C.Y.: Representation of switching circuits by binary-decision programs. The Bell System Technical Journal 38(4), 985–999 (1959)

41. Markov, A.A.: Introduction to Coding Theory. Nauka, Moscow (1982) (in Russian)

42. Martelli, A., Montanari, U.: Optimizing decision trees through heuristically guided search. Commun. ACM 21, 1025–1039 (1978)

43. Moret, B.E., Gonzalez, R.C., Thomson, M.: activity of a variable and its relation to decision trees. ACM Trans. Program. Lang. Syst. 2, 580–595 (1980)

44. Moret, B.E.: Decision trees and diagrams. ACM Comput. Surv. 14, 593–623 (1982)

45. Moshkov, M.: On conditional tests. Soviet Physics Doklady 27, 528 (1982)

46. Moshkov, M.: Conditional tests. Problemy Kybernetiki 40, 131–170 (1983) (in Russian)

47. Moshkov, M.: Decision Trees.Theory and Appliucations Nizhni Novgorod University Publishers (1994) (in Russian)

48. Moshkov, M.: About the depth of decision trees computing Boolean functions. Fundam. Inf. 22(3), 203–215 (1995)

49. Moshkov, M.: Some bounds on minimal decision tree depth. Fundam. Inf. 27(2/3), 197–203 (1996)

50. Moshkov, M.J.: Time complexity of decision trees. In: Peters, J., Skowron, A. (eds.) Transactions on Rough Sets III. LNCS, vol. 3400, pp. 244–459. Springer, Heidelberg (2005)

51. Moshkov, M., Chikalov, I.: On the average depth of decision trees over information systems. In: Proceedings of the Fourth European Congress on Intelligent Techniques and Soft Computing, Aachen, Germany, vol. 1, pp. 220–222 (1996)

52. Moshkov, M., Chikalov, I.: Upper bound on average depth of decision trees over information systems. In: Proceedings of the Fourth International Workshop on Rough Sets, Fuzzy Sets and Machine Discovery, Tokyo, Japan, pp. 139–141 (1996)

53. Moshkov, M., Chikalov, I.: Bounds on average depth of decision trees. In: Proceedings of the Fifth European Congress on Intelligent Techniques and Soft Computing, Aachen, Germany, pp. 226–230 (1997)

54. Moshkov, M., Chikalov, I.: Bounds on average weighted depth of decision trees. Fundam. Inf. 31(2), 145–156 (1997)

55. Moshkov, M., Chikalov, I.: On algorithm for constructing of decision trees with minimal depth. Fundam. Inf. 41, 295–299 (2000)

56. Moshkov, M., Chikalov, I.: Consecutive optimization of decision trees concerning various complexity measures. Fundam. Inf. 61(2), 87–96 (2003)

57. Nguen, H.S., Nguen, S.H.: From optimal hyperplanes to optimal decision trees: rough set and boolean reasoning approaches. In: Proceedings of the Fourth International Workshop on Rough Sets, Fuzzy Sets and Machine Discovery, Tokyo, Japan, pp. 82–88 (1996)

58. Nguen, H.S., Nguen, S.H.: Discretization methods in data mining. In: Polkowski, L., Skowron, A. (eds.) Rough Sets in Knowledge Discovery 1. Methodology and Applications. Studies in Fuzziness and Soft Computing, vol. 18, pp. 451–482. Physica-Verlag, Heidelberg (1998)

59. Okolnishnikova, E.A.: Lower bounds for branching programs computing characteristic functions of binary codes. Metody Discretnogo Analiza 51, 61–83 (1991)

60. Pattipati, K.R., Dontamsetty, M.: On a generalized test sequencing problem. IEEE Transactions on Systems, Man, and Cybernetics 22(2), 392–396 (1992)

61. Pawlak, Z.: Information Systems – Theoretical Foundations. PWN, Warsaw (1981) (in Polish)

62. Pawlak, Z.: Rough Sets: Theoretical Aspects of Reasoning about Data. Kluwer Academic Publishers, Norwell (1992)

63. Picard, C.F.: Theorie des Questionnaries. Gauthier-Villars, Paris (1965) (in French)

64. Pollack, S.L.: Conversion of limited-entry decision tables to computer programs. Commun. ACM 8, 677–682 (1965)

65. Pollack, S.L.: Decision Tables: Theory and Practice. John Wiley & Sons, Chichester (1971)

66. Ponzio, S.: Restricted branching programs and hardware verification. Ph.D. thesis, Massachusetts Institute of Technology (1995)

67. Post, E.: Introduction to a general theory of elementary propositions. Amer. J. Math. 43, 163–185 (1921)

68. Post, E.: The two-valued iterative systems of mathematical logic. Annals of Mathematical Studies 5, 1–122 (1941)

69. Preparata, F.P., Shamos, M.I.: Computational Geometry: an Introduction. Springer, Heidelberg (1985)

70. Quinlan, J.R.: Induction of decision trees. Mach. Learn. 1, 81–106 (1986)

71. Quinlan, J.R.: C4.5: Programs for Machine Learning (Morgan Kaufmann Series in Machine Learning), 1st edn. Morgan Kaufmann, San Francisco (1993)

72. Redkin, N.P.: Reliability and Diagnostics of Circuits. Izd-vo MGU, Moscow (1992) (in Russian)

73. Reinwald, L.T., Soland, R.M.: Conversion of limited-entry decision tables to optimal computer programs I: Minimum average processing time. J. ACM 13, 339–358 (1966)

74. Rosten, E., Drummond, T.: Fusing points and lines for high performance tracking. In: IEEE International Conference on Computer Vision, vol. 2, pp. 1508–1511 (2005)

75. Rosten, E., Drummond, T.: Machine learning for high-speed corner detection. In: European Conference on Computer Vision, vol. 1, pp. 430–443 (2006)

76. Schumacher, H., Sevcik, K.C.: The synthetic approach to decision table conversion. Commun. ACM 19, 343–351 (1976)

77. Shannon, C.E.: A mathematical theory of communication. Bell System Technical Journal 27, 379–423, 623–656 (1948)

78. Shwayder, K.: Conversion of limited-entry decision tables to computer programs a proposed modification to Pollack's algorithm. Commun. ACM 14, 69–73 (1971)

79. Skowron, A., Rauszer, C.: The discernibility matrices and functions in information systems. In: Slowinski, R. (ed.) Intelligent Decision Support. Handbook of Applications and Advances of the Rough Set Theory, pp. 331–362. Kluwer Academic Publishers, Dordrecht (1992)

80. Slagle, J.R.: An efficient algorithm for finding certain minimum-cost procedures for making binary decisions. J. ACM 11, 253–264 (1964)

81. Solovyev, N.A.: Tests (Theory, Construction, Applications). Nauka, Novosibirsk (1978) (in Russian)

82. Thayse, A., Davio, M., Deschamps, J.P.: Optimization of multivalued decision algorithms. In: Proceedings of the Eighth International Symposium on Multiple-Valued Logic, MVL 1978, pp. 171–178. IEEE Computer Society Press, Los Alamitos (1978)

83. Wegener, I.: The Complexity of Boolean Functions. John Wiley & Sons, B.G. Teubner, Stuttgart (1987)

84. Wegener, I.: On the complexity of branching programs and decision trees for clique functions. J. ACM 35, 461–471 (1988)

85. Zák, S.: An exponential lower bound for one-time-only branching programs. In: Proceedings of the Mathematical Foundations of Computer Science, pp. 562–566. Springer, London (1984)

Index

$M_\Psi(z)$ parameter   7, 37–39, 80

algorithms
  $\mathcal{A}$,   12, 62–69, 83–84
  greedy,   70–71

branching program,   10, 59
  implementing a function,   60
  read-once,   10, 59

complexity measure,   7
  average depth,   7, 8, 18
  average weighted depth,   7, 8
  depth,   7, 10, 42

decision table,   5
  partition,   50
decision tree,   5
  complete path in,   6
  implementing Boolean
    function,   9, 42, 72–74
  irredundant,   11, 63
  optimal,   7
  solving the problem,   6

entropy of probability distribution,   8

information system,   4
  a system of equations over,   13, 80
  restricted,   80–82

probability distribution,   5
  Q-restricted,   12, 69, 83
problem over information system,   4
  decomposition,   9, 26–36
  diagnostic,   5, 8, 16
  with a complete set of
    attributes,   8, 16

separable subtables,   6
  a graph of,   64–66
  compatible set of,   50
  terminal,   6
Shannon-type functions
  $\mathcal{D}_U(n)$,   14, 83
  $\mathcal{G}_B(n)$,   10, 42
  $\mathcal{H}_B(n)$,   10, 42–58
  $\mathcal{S}_U(n)$,   14, 83

threshold function,   48

weight function,   4, 80

CPSIA information can be obtained at www.ICGtesting.com
Printed in the USA
LVOW012016021111

253240LV00001B/4/P